Reginald Harrison

Lectures on the Surgical Disorders of the Urinary Organs

Delivered at the Liverpool Royal Infirmary. Second Edition

Reginald Harrison

Lectures on the Surgical Disorders of the Urinary Organs
Delivered at the Liverpool Royal Infirmary. Second Edition

ISBN/EAN: 9783337269890

Printed in Europe, USA, Canada, Australia, Japan

Cover: Foto ©berggeist007 / pixelio.de

More available books at **www.hansebooks.com**

LECTURES

ON

THE SURGICAL DISORDERS

OF THE

URINARY ORGANS.

DELIVERED AT THE LIVERPOOL ROYAL INFIRMARY.

BY

REGINALD HARRISON, F.R.C.S.,

SURGEON TO THE INFIRMARY;
FORMERLY LECTURER ON ANATOMY AND SURGERY AT THE SCHOOL OF MEDICINE; AND
SURGEON TO THE LIVERPOOL NORTHERN HOSPITAL.

SECOND EDITION, CONSIDERABLY ENLARGED.

LONDON:
J. & A. CHURCHILL, NEW BURLINGTON STREET.

LIVERPOOL:
ADAM HOLDEN, CHURCH STREET.

1880.

LIVERPOOL:
D. MARPLES AND CO., LIMITED,
LORD STREET.

TO

THE STUDENTS,

PAST AND PRESENT,

OF

THE LIVERPOOL ROYAL INFIRMARY,

THESE LECTURES ARE

𝔇𝔢𝔡𝔦𝔠𝔞𝔱𝔢𝔡

BY THEIR SINCERE FRIEND,

THE AUTHOR.

PREFACE TO THE SECOND EDITION.

I HAVE endeavoured to make a second edition of these Lectures more worthy of the favour that was accorded to the previous one.

With this object, not only have I added to the original Lectures references to many improvements in practice, but my remarks are considerably extended for the purpose of embracing the larger field of the Surgery of the Urinary Organs.

I desire to acknowledge the great assistance received from many professional friends in furnishing illustrations, which enabled me without difficulty to treat my subject in a more systematic manner than could otherwise have been undertaken in a clinical course.

To Mr. F. T. PAUL I am again indebted for much valuable aid rendered in a variety of ways.

Nearly every pathological specimen referred to in these pages will be found in the Museum of the Liverpool Royal Infirmary; and I would take this opportunity of expressing my obligations to the Chairman and the Committee of that Institution for the maintenance of so important a means as is provided by this collection for promoting the advancement of Practical Medicine and Surgery.

38, RODNEY STREET, LIVERPOOL,
NOVEMBER, 1880.

CONTENTS.

FIRST LECTURE.

Preliminary Remarks — Definition of Stricture — Causes — Gleet — The Use of Injections — Position of Strictures — The Endoscope — Classification — Varieties of Stricture — — 1

SECOND LECTURE.

The Surgical Anatomy of the Urethra — Spasm — The Dimensions of the Urethra — Otis's Views — The Curvature of the Urethra — The Relation of the Urethra to the Rectum — Attachments of Fasciæ — Opening of the Seminal Ducts — 19

THIRD LECTURE.

Symptoms of Stricture — Granular Urethritis — Consequences of Stricture on the Genito-Urinary Organs — Nervous and Spasmodic Affections simulating Stricture — — — — — 33

FOURTH LECTURE.

Examination of the Urine — — — — — — — — — — 46

FIFTH LECTURE.

Treatment of Stricture — Gradual Dilatation — Instruments Employed — The Filiform Bougie — Anæsthetics — Continuous Dilatation — — — — — — — — — — — — 65

CONTENTS.

SIXTH LECTURE.

Urethral Fever—Suppression of Urine—Hæmorrhage from the Urethra—False Passages - - - - - - - - - - 89

SEVENTH LECTURE.

Retention of Urine — Catheterism — Impassable Stricture—Aspiration of the Bladder—Tapping—Cock's Operation—Forcible Catheterism - - - - - - - - - - 101

EIGHTH LECTURE.

Internal Urethrotomy—Selection of Cases—Otis's Views—Various Methods of Performing Internal Urethrotomy—The Use of Oval Bougies—Holt's Operation - - - - - 112

NINTH LECTURE.

External Urethrotomy—Syme's Operation—Selection of Cases—External Urethrotomy with a Guide—Without a Guide—Wheelhouse's Operation—Subcutaneous Urethrotomy - - 125

TENTH LECTURE.

Syphilitic Strictures—Nunn's and Bell's Views—Chancre of the Meatus—Cases of Stricture Complicated with Syphilis—Treatment - - - - - - - - - - - - - 136

ELEVENTH LECTURE.

Consequences of Stricture—Urethral Abscess—Extravasation of Urine - - - - - - - - - - - - - - 145

TWELFTH LECTURE.

Injuries to the Urethra—Contusion—Rupture of the Urethra—Cases—Treatment—Longitudinal Wounds of the Urethra - 157

THIRTEENTH LECTURE.

Perinæal Fistulæ and their Treatment - - 166

FOURTEENTH LECTURE.

Foreign Bodies in the Urethra and Bladder—Action of Urethra—Illustrative Cases—Use of the Lithotrite as an Extractor—Foreign Bodies in the Female Bladder - - - - - - - 179

FIFTEENTH LECTURE.

Irritable Bladder - - - - - - - - - - - - - 190

SIXTEENTH LECTURE.

Hypertrophy of the Prostate—Retention of Urine—Circumstances under which Retention occurs—Treatment—Incontinence of Urine—Formation of Calculi—Operative Treatment of Enlarged Prostate - - - - - - - - 209

SEVENTEENTH LECTURE.

Inflammation of the Bladder—Atony - - - - - - - - 225

EIGHTEENTH LECTURE.

On the Formation and Physical Constitution of Urinary Calculi 239

NINETEENTH LECTURE.

Spontaneous Fracture of Calculi—Varieties in their Composition and Shape - - - - - - - - - - - - - 248

TWENTIETH LECTURE.

Calculous Disorders—Stone in the Kidney—Renal Colic—Calculi impacted in the Ureter - - - - - - - - 262

TWENTY-FIRST LECTURE.

Symptoms of Stone in the Bladder—Sounding—The Microphone—Sources of Error in Sounding - - - - - - - 270

TWENTY-SECOND LECTURE.

Treatment of Calculous Disorders—Lithotomy - - - - - 285

TWENTY-THIRD LECTURE.

Lithotrity—Litholapaxy—Stone in Females—Treatment - - 303

TWENTY-FOURTH LECTURE.

Injuries to the Bladder - - - - - - - - - - - - - 315

TWENTY-FIFTH LECTURE.

Injuries to the Ureter and Kidney—Hernia of the Kidney - - 328

TWENTH-SIXTH LECTURE.

Surgery of the Kidney - - - - - - - - - - - - - 337

TWENTY-SEVENTH LECTURE.

Tumours of the Bladder and Prostate—Scirrhus and Medullary Cancer—Non-malignant Tumours—Villous Growths—Palpation of the Rectum - - - - - - - - - - - - 351

TWENTY-EIGHTH LECTURE.

Ulcerations of the Bladder—Tubercular and Cancerous Ulcerations—Perforating Ulcerations—Colotomy—Sloughing of the Bladder - - - - - - - - - - - - - - - 363

TWENTY-NINTH LECTURE.

Circumcision—Deformities of the Frænum and of the Meatus—
  Hypospadias—Epispadias—Patent Urachus—Absence of
  Bladder—Amputation of the Penis 373

THIRTIETH LECTURE.

Varicocele—Treatment; Palliative and Radical . . 388

LIST OF ILLUSTRATIONS.

LITHOGRAPHS.

		PAGE
Temperature Chart		92
A.	Grass-head in the Bladder	184
B.	Enlarged Prostate	215
C.	False Passages through the Prostate	217
D.	Calculi concealed by a large Prostate	219
E.	Calculi of unusual shape	257
F.	Calculus impacted in the Ureter	268
G.	Calculus in a Soft Envelope	282
H.	Villous Growth within the Bladder	356

WOODCUTS.

FIG.		PAGE
1.	Diagram of Urethra	11
2.	Urethral Irrigation	12
3.	Filiform Bougie adapted for an Eccentric Stricture	16
4.	Cast of Urethra	24
5.	Side View of Bladder and Urethra (from Gray)	26
6.	Side View of Pelvic Fasciæ	29
7.	Healthy and Granular Urethra	35
8.	Catheter impeded by the Triangular Ligament	68
9.	Bougie-à-Boule	70
10.	Do. do.	70
11.	Do. do.	70
12.	Conical Bougie	71
13.	Tunneled Bougie and Catheter	75
14.	Catheter Gauge	78
15.	Watson's Steel Probe-pointed Catheter	103
16.	Probe-ended Knife	116
17.	Watson's Urethrotome	116

FIG.		PAGE
18.	Author's Urethrotome, with Knife closed	118
19.	Do. do. with Knife projected	118
20.	Wheelhouse's Perineal Staff	129
21.	Teale's Probe-gorget	130
22.	Wheelhouse's Staff introduced	131
23.	Probe-gorget do.	133
24.	Diagram illustrating Treatment of Urethral Fistula	173
25.	An Incrusted Needle removed from the Urethra	181
26.	Phosphatic Calculus	182
27.	Needle removed from the Perinæum	184
28.	Pencil Case removed from the Urethra	186
29.	Robert and Collin's Extractor	187
30.	Diagram illustrating Reflex Irritability of the Bladder	196
31.	French Gum-elastic Prostatic Catheter	215
32.	Apparatus for emptying the Bladder	229
33.	Pessary Catheter	234
34.	Relative Anatomy of the Urethra	289
35.	Browne's Dilatable Tampon	295
36.	Tiemann's Bladder Shield	383
37.	Keetley's Support for Varicocele	391
38.	Morgan's Suspensory Bandage	391

FIRST LECTURE.

Preliminary Remarks—Definition of Stricture—Causes— Gleet—The Use of Injections—Position of Strictures—The Endoscope—Classification—Varieties of Stricture.

Gentlemen,—The practice of this Infirmary affords you abundant opportunities for observing the surgical disorders of the urinary organs; and of these, cases of stricture of the urethra, and the complications arising from it, form no inconsiderable proportion.

As I purpose devoting the first portion of this course to the discussion of stricture and its treatment, I shall proceed to make some general remarks in reference to this affection.

Associated as I have been for some years with two of the hospitals in this town deriving a large number of their patients from the seafaring population connected with the port, my observation leads me to believe that amongst this class of the community stricture is a common disorder. And that it should be so is not surprising. Gonorrhœa, contracted on shore in the debauch that frequently precedes a vessel's departure for some foreign port, breaks out two or three days afterwards. Treatment, except in ships carrying passengers—and sometimes even in these—

is usually conducted by the captain or the mate, and not always, as might be anticipated, with advantage to the patient. The old notion that every disorder consequent on promiscuous intercourse is "venereal," and must be treated by mercury, still prevails, and large doses of calomel, until profuse salivation is produced, is not rarely the only remedy administered for a gonorrhœal discharge. Some of the worst cases of stricture that I have seen have been occasioned, under similar circumstances, by resort to the most primitive means for the relief of retention of urine. In the absence of catheters from the ship's medicine chest, or still more frequently, as I have found, on account of their rottenness, I have known instances where the wire from a soda-water bottle and an iron skewer have each done duty in "forcing" a stricture. A bougie made from an old clock-pendulum was recently shown me by a sailor, as an instrument modelled by himself with which he had successfully combated an obstinate stricture, that had previously resisted the attacks made on it with a gum-elastic catheter. It is not a long time ago that a man was admitted into my ward with retention, and a badly lacerated urethra, as a consequence of an attempt on the part of the mate of his ship to reach the bladder by the aid of a pointed piece of wood, roughly adapted to the shape of a bougie. Here the operator was more than professionally interested, inasmuch as he had occasioned the retention by kicking the patient behind the scrotum. The most remarkable piece of ingenuity of the kind that I can remember was where, after a patient had endured the unspeakable agonies of retention for over three days, an endeavour had been made to introduce through the urethra a piece of gas-piping, which had been devised, *in extremis*, for the purpose by the engineer of the ship. Unfortunately, however, this failed to effect its object. When I saw him, immediately on his arrival (the captain of the ship signalling the circumstance when off Holyhead), I found the urethra much lacerated, and it was with considerable difficulty that I introduced a catheter, and removed a large quantity of the most fetid urine imaginable. Relief, however, came too late, the man dying shortly after his arrival, with convulsions and urinary poisoning. Though deploring that persons should by circumstances be placed in such unfortunate positions, I allude to these cases for the purpose of showing you that your field for observation here, in this department of surgery alone, is by no means restricted to what I may call routine, or even to the freaks of nature or the ravages of disease.

In undertaking to speak about the treatment of stricture, I am conscious that the subject is a well-worn one. Still, with all our plans of treatment, we have not arrived at anything like uniformity in practice; and as this is only to be attained by taking the sum of our respective experiences, I feel less hesitation in bringing under your notice some conclusions which I have been enabled to gather from both public and private sources. These considerations I hope to place before you during that portion of the year in which it will be my duty to conduct the clinical lectures in surgery. I shall endeavour to give you this information in a more systematic manner than is attempted in lectures specially devoted to clinical purposes; but as my wards are usually well supplied with cases necessary for illustrating such a course as that which will be commenced to-day, I anticipate no difficulty, so far as clinical material is concerned, in giving effect to my wishes.

In using the term "stricture" I reserve it, as Sir Henry Thompson suggests in his practical work on *Diseases of the Urinary Organs*, for one kind of stricture—namely, organic stricture. "Spasm" and "inflammation" are conditions which may be superadded, but they do not constitute stricture in the acceptation of the term which is now generally adopted. By stricture then I must be understood to refer to organic stricture, where the impediment to micturition is occasioned by some structural change in the wall of the urethra. The causes of stricture are various. The greater proportion of patients attribute their misfortune, directly or indirectly, to previous attacks of gonorrhœa. Those who do so *directly* are disposed to regard the stricture as the natural consequence of their previous mishap. Those who do so *indirectly*, usually have something to say about the treatment employed and its bearing upon the subsequent formation of a stricture. It is worth our while, for a moment, to analyse the statements made by this latter class, with the view of ascertaining how far their allegations hold good. "I was almost cured of my gonorrhœa, only a very slight discharge remaining, which I thought would go away of itself," is the statement of the patient who is convicted of his own indiscretion in having allowed things to go on from bad to worse. Others, again, seek refuge in referring their misfortune to the improper advice they have received. "I was told that it was only a gleet, due to weakness, which would go away by iron, tonics, and cold baths." Here we have illustrations of gleet terminating in stricture.

Now, it is well for you, once for all, to understand that a gleet is not a disorder which is disposed to go away of itself; on the contrary, it requires careful and well-considered treatment; and if it does not receive this—that is to say, if it is clumsily dealt with, or not dealt with at all—it most probably ends in the formation of a stricture.

A gleet is to be regarded as indicative of the early formation of stricture. Nay, further, you will not do wrongly in regarding a gleet as the stage in the stricture-forming process when, by your treatment, you can promise your patient to restore his urethra to its normal condition. When a stricture is once allowed to become cicatricial in its character, you may palliate or adapt, but you can no more *restore* his urethra than you can, by dissection or any other process, remove a scar from his skin. You may moderate the inconveniences of a scar, but you cannot obliterate it. Let not, then, the curable stage of stricture pass by; at all events, let the onus of doing so rest with your patient, and not with yourself.

Again, it is very common to hear patients attribute their strictures to the use of injections in the treatment of their gonorrhœas. A considerable amount of prejudice exists in the public mind in reference to the use of these applications. Patients not unfrequently say, when consulting you about a gonorrhœa, "Do not order me an injection, as I understand these remedies often occasion stricture." Is there any truth in such an allegation? Assuredly not, presuming, of course, injections are judiciously prescribed and properly used.

Let me remind you that the cure of gonorrhœa by specifics is essentially one on the principle of injection. For how do the drugs that act specifically on the urethra effect their purpose? How do we explain the action of copaiba, oil of sandal-wood, creasote, and certain terebinthinates, in the cure of gonorrhœa? Do not all these drugs exercise their therapeutic virtues by certain of their constituents, for the most part demonstrable, being conveyed by the urine to the site of the disorder? What is this but a cure by injection, or, to be etymologically correct, ejection? It is the urine of the patient that conveys the specific to the disease, just as the rose-water in your injection does the sulphate of zinc, or other astringent.*

* Somewhat similar results appear to follow inhalation. Taking the idea from Professor Dittel's paper on the benefit obtained in cases of pyelitis and catarrh of the kidney-tubes by inhaling certain essential oils, Zeissl tried it in purulent urethritis. The case selected was a male with a copious discharge, and he was caused to inhale the vapour of rectified oil of turpentine, morning and evening, for a quarter of an hour at a time. On the second day the urine betrayed the odour which is likened to the smell of violets, and is usual in that secreted by patients taking the oleo-resins. The inhalations caused no inconvenience, and were steadily continued for twenty-five days, at the end of which

It is the abuse of injecting that is open to animadversion. Injections, in the treatment of gonorrhœa, do harm only when, by reason of their composition or strength, they act as *irritants* to the mucous membrane.

In the ordering of urethral injections, there are two rules which should be followed :—1. Do not strain the urethra by the *quantity* of injection used. 2. Do not pain the urethra by the *quality* of the injection. A teaspoonful of fluid *put* into the urethra frequently is better than a tablespoonful *forced* in three times a day. This is a point upon which I have long insisted. In prescribing injections you should feel your way, adding to their strength according to circumstances. Some persons, it is well known, are far more sensitive to the action of remedies than others; and this applies equally to the urethra—" The temper of the urethra varies as much as the temper of the mind."* An injection appropriate in strength to a first gonorrhœa, is like the proverbial drop of rain on a duck's back in the case of the *habitué*. I remember ordering a patient of this class an injection, well known as "the four sulphates." It cured him effectually, and without pain. A friend, hearing of the success, borrowed the prescription, and, without proper advice, used it. The consequences were, an

---

time the pus had entirely disappeared from the urine. The experiment was repeated in a second case, with like results.—*London Medical Record*, vol. i., p. 361.

A curious effect of what I believe to have been due to inhalation was brought before the notice of the Liverpool Medical Institution, in 1866, by Mr. T. Shadford Walker. The whole crew of a ship, carrying a cargo of turpentine, had suffered during the voyage from hæmaturia; in one case with a fatal result.

* Brodie, *On Diseases of the Urinary Organs*, p. 50.

acute attack of cystitis, and a subsequent stricture. Surely, it is only to the foolhardiness of the sufferer that such an unfortunate result is to be attributed.

And I would here remark, that I have seen a great deal of damage done, and suffering occasioned, by the use of some of the nostrum injections advertised throughout the country as "infallible cures" and "preventives." Many of them contain the ordinary astringents applicable to the urethra, but in a very potent form. I caution you therefore against sanctioning their use.

These observations have been made with the view of showing that it is only by their improper use that injections are open to the charge of occasioning stricture. If they are prescribed in accordance with the rules I have given, you will never have cause to regret their employment.

Then, again, we have strictures resulting from the healing of sores, of wounds, and of lacerations of the urethra; in the last-named we include what are described as traumatic strictures. Let me take a few illustrations. A patient has a sore on his glans penis, involving the meatus. When the sore heals a cicatrix is left. All scars or cicatrices are prone to contract, and thus cause a narrowing of the urethral orifice. This condition was well illustrated in a patient where the same state of things was brought about by an improperly performed operation of circumcision, a portion of the glans penis having been removed as well as the prepuce. When the sore healed, the cicatrix contracted, and the patient presented himself here with a tight stricture of the meatus requiring division. Blows on the perinæum are a frequent cause of stricture amongst our sailor population. A man falls from aloft, across a spar or a rope, and ruptures his urethra. If the patient recover from the immediate effects of the injury, it is with his urethra scarred. This is the worst variety of stricture—traumatic,—a form of the disorder more obstinate to deal with than any other. Traumatic stricture may be caused by violence from within, as in crushing of the pelvis, when the fractured pubic bone pierces or lacerates the urethra. Here we have, for the most part, jagged and uneven wounds, not rarely complicated with extravasation of urine—a condition of things unfavourable to accurate or kindly repair. Injuries occasioned in this way are invariably followed by a dense stricture, and in this respect are a contrast to the clean-cut wounds of lithotomy, where the occurrence of stricture is exceedingly rare. This observation has an important practical bearing, and will again be alluded to in considering the treatment of injuries to the urethra.

I need not give further illustrations of the causes of stricture. Anything that is capable of structurally altering the wall of the urethra, by scarring it, as in wounds, however inflicted, or indurating it, as by the persistence of inflammatory action, is liable to be followed by a stricture; and any portion of the urethra (excepting the prostatic, which has its own peculiar obstruction,) may be so strictured.

The greater proportion of strictures occur at the

sub-pubic curvature of the urethra; here for their detection and treatment we are dependent upon various kinds of instruments known as bougies; consequently it is of the first importance that we should accustom ourselves to their use, in order that we may diagnose correctly and treat skilfully.

Various explanations have been offered of the frequency of stricture in the position referred to. It is quite obvious why traumatic stricture should most frequently occur here, as the urethra is violently dragged, by the force applied, from where it is fixed by its connection with the triangular ligament. It is at first sight less obvious in reference to stricture resulting from chronic urethritis. I think the explanation offered by Mr. A. P. Gould is probably the correct one.[*] By reference to the diagram accompanying Mr. Gould's paper, it will be seen that the portion of the urethra which is more frequently strictured than any other is horizontal, and here it is that morbid secretions are apt to collect and to cause a sufficient amount of inflammatory action to lead to plastic exudation within and around the walls of the urethra. I believe that this is the correct explanation of a generally admitted fact; it is also suggestive as to what may be done in cases such as these to prevent the formation of stricture. For some time past I have been employing, in the treatment of obstinate gleety discharges, the thorough irrigation of the deeper portion of the urethra, by means of a soft catheter and slightly

[*] *Why is Organic Stricture most common in the Bulbous Portion of the Urethra?* By A. Pierce Gould. *Lancet*, Dec. 8, 1877.

astringent solutions, with most satisfactory results. Sufficient importance is not, I believe, attached to the thorough irrigation of the bulbous portion of the

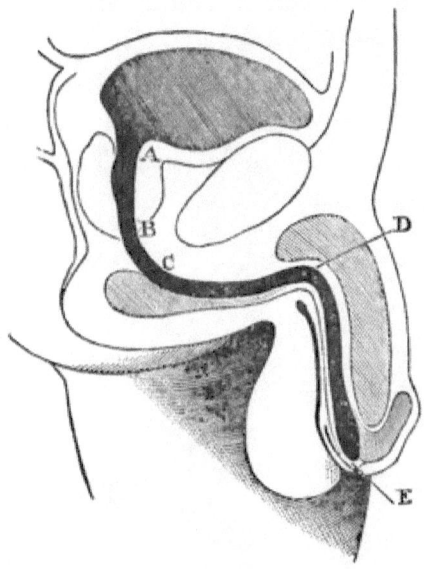

Fig. 1.

A—C. Prostatic and Membranous Urethra.
C—D. Bulbous   do.
D—E. Penile Urethra.

urethra in the treatment of gleet. The fact that this locality is by far the most frequent position of stricture; that stricture is almost always preceded by gleet; and that the ordinary mode of using injections, as practised for the treatment of gonorrhœa, is utterly valueless in gleet; all point to a conclusion which experience justifies, that to cure a gleet and prevent a stricture, the part affected must be completely brought under the influence of medication. This can only be efficiently done by the use of a suitable urethral douche, by means of which the urethra may be thoroughly cleansed from one end to the other twice a day, or oftener if necessary, employing for this purpose as a lotion several ounces of tepid water, containing some sulpho-carbolate of zinc or other similar agent. I have found the practice followed by great success."

The apparatus † I make use of is shown in the following sketch :—

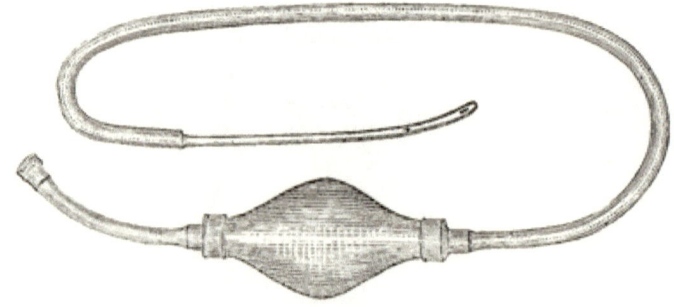

Fig. 2.

It consists of a small Higginson's syringe, upon which is fitted an india-rubber catheter. By means of this appliance the patient is directed to *douche* his urethra, not using, as with the ordinary glass syringe, about a teaspoonful of fluid, but half a pint of the necessary astringent. In old cases of gleet, where there is a granular bulb, the conventional mode of injecting with the glass syringe is of about as much service as a gargle is to the ulcerated throat of a patient who contents himself with taking half a wine-

* *Lancet*, May 15th, 1880.
† It can be procured from Messrs. Symes, Hardman Street, Liverpool.

glassful of the fluid into his mouth and then spitting it out, and repeating the performance thrice daily.

Less frequently, we find strictures occurring at the meatus of the urethra, and within two and a half inches of it; these are generally occasioned by the puckering resulting from the healing of venereal sores. You will remember the very marked illustration of this variety which we had recently, where the last inch of the urethra was strictured by a mass of cicatricial material, which had resulted from the filling up of an extensive phagedænic sore.

It will be proper, here, to remind you that, for the exploration of the urethra, a special speculum, or, as it is called, an endoscope, has been devised. The most improved instrument is Desormeaux's, as modified by Dr. Cruise, of Dublin, where a strong artificial light is reflected along a urethral speculum. A tolerably extended use of this instrument has convinced me that its utility is limited to certain cases of granular urethritis, which resist the ordinary methods of applying topical agents.* In the diagnosis and treatment of stricture, it furnishes little or no assistance.

More recently the galvanic light has been used for the purpose, but so far the results do not appear to me to be any more encouraging than those arrived at by the ordinary endoscope. †

There are two forms of stricture, the mucous and sub-mucous, which, though often concurrent, yet

---

* *The Endoscope in Diseases of the Urethra and Rectum*, by Reginald Harrison.—*British Medical Journal*, 1868.

† Sir Henry Thompson's *Lecture on the Nietze-Leiter Endoscope.*—*Lancet.* Dec. 6th, 1829.

generally have an independent existence; and as, bearing upon the subject of treatment, their recognition is of importance, I refer to them here.

By the mucous stricture we understand that form of the affection in which the impediment is limited to the lining membrane of the urethra. This impediment consists, for the most part, of the puckerings caused by the healing of ulcers—the caruncles of the old writers,—and adhesions of the membrane forming those obstacles which have received the name of valves or bridles.

In the sub-mucous variety, the obstruction is in the tissues outside the mucous membrane. So structurally unimplicated may this membrane be, that, on its removal from the indurated tissue beneath, it is found unaltered in its dimensions. This I have verified after death, by dissection. During life, this position of the stricture explains the great success that follows Holt's operation in certain cases, the effect of such an operation being to break up the stricture without any further damage to the mucous membrane than that of stretching it to its original dimensions.*

The following classification of strictures (after Dittel) † may be found useful for reference. In this

---

* Benjamin Bell, in his work on Gonorrhœa, published in 1793, recognised the distinction. "In the more fixed kinds of obstruction proceeding from gonorrhœa, the diameter of the urethra is lessened in two different ways. For the most part, it is diminished by a thickening taking place at some particular point in the membrane of the passage itself. . . At other times, the urethra is drawn together, or contracted as if a cord were tied round it, without any disease being perceptible."

† *Die Stricturen der Harnrohre*, von Professor Dr. Leopold Dittel.

are included strictures due to new growth, excluding heteroplastic growths, such as malignant and other tumours pressing upon the canal, the new growth being connective tissue, which always has a tendency to contract when not adequately resisted.

First.—Free (inside the canal), including warts, valves, and bridles.

Second.—In the walls, including traumatic and ulcer cicatrices.

Third.—Outside and around the mucous membrane, including peri-urethral callus, as—

> Ring stricture (short).
> Nodular stricture.
> Diffuse stricture.

It not unfrequently happens that the urethra is strictured at more than one point. Hunter gives an instance where a urethra contained six strictures; and even more than these have been found by Lallemand and other French writers. You will, however, seldom meet with instances where the number of strictures exceeds three; two are not uncommon. Where there are multiple strictures, and the anterior one is exceedingly tight, the difficulty of passing those beyond is much increased, inasmuch as the free manipulation of the bougie is interfered with by the tightness with which it is grasped by the anterior obstruction. In such instances it is often necessary to deal with the anterior stricture independently, either by dilatation or incision, until it is sufficiently relaxed to allow of the instrument being passed with freedom on to the face of the deeper obstruction. You have had several illustrations lately of the advantages to be derived in multiple stricture from this plan of proceeding.

It must be remembered, especially in connection with the introduction of instruments, that the course of the urethra is often deviated by reason of the cicatricial material constituting the stricture being irregularly deposited. Hence we have what are called concentric and excentric strictures. In the former the passage, though contracted, still remains central; whilst in the latter, which includes the large majority of strictures, the canal is pushed to one side, and thus rendered not only more or less impervious, but also tortuous.

I frequently demonstrate this when using the fine whalebone bougies in cases of advanced stricture. In passing them, if I find any hitch to their progress when the stricture is reached, I withdraw them and bend their tips to an angle, as represented in the accompanying sketch :—

Fig. 3.

On re-introducing them, and giving them a spiral movement on approaching the stricture, I usually find no difficulty in then carrying them on into the bladder, the explanation being, that the axis of the canal is not central. You have, in fact, to seek the opening through the stricture in a corner or angle of the canal, and not, as we generally expect to find it, in the middle. The small gum-elastic filiform bougies adapt themselves to this condition of things, and hence have an advantage over the unyielding metallic instruments. Ignorance of this in using metallic instruments frequently leads, I believe, to the formation of false passages, force being substituted for the *tactus eruditus* which is alone necessary. Benjamin Bell, and, more recently, Leroy d'Etiolles and Gouley, of New York, have all in their writings insisted upon the due recognition of this very important practical point.

Just as there are differences in the form and position of strictures, so are there differences in what I may call their temper. One is indolent, whilst another is irritable; a third is contractile, and a fourth hæmorrhagic. I mention them now to prevent repetition when I come to speak of treatment, and for the purpose of reminding you at once that much discrimination and experience are required, not only in the carrying out, but still more in the selection, of the means appropriate to each case.

The irritable stricture, as the name implies, is more or less intolerant of all instrumental interference, such as the passing of bougies or of catheters. It is usually accompanied with muscular spasm, which adds considerably to the distress of the patient. Hence this stricture is not amenable to treatment by gradual dilatation, as each passing of an instrument is commonly attended by some urethral distress, such as retention, the discharge of blood, or a rigor, followed by varying degrees of febrile excitement. These strictures are best met by some proceeding in which, under an anæsthetic, the contracted part is at once brought up to the normal calibre of the urethra; of these, as I shall subsequently state, Holt's, or the immediate method, is perhaps the best; at all events, I have found it so; and if in these irritable strictures, intolerant of the prolonged presence or otherwise of an instrument in the urethra, I employ dilatation, it is merely for the purpose of bringing the contracted urethra to a size capable of receiving a dilator which such an operation requires. These cases are not suitable, as a rule, for any form of urethrotomy, or cutting operation, as they will be found to furnish the instances where such a proceeding has been followed by a fatal result.

The contractile stricture is one that you can dilate, but it does not retain, even for twenty-four hours, any of the additional calibre you may have given it by the gradual introduction of bougies in successive sizes. It simply speedily falls back to its original narrowed dimensions. These are the strictures which are generally benefited by section—that is to say, by external or internal urethrotomy. This form of stricture has been described with singular accuracy and clinical precision by the late Professor Syme, who, by the operation he perfected, brought it within the range, if not always of cure, at all events of such an amount of relief as to render it readily manageable.

The terms "indolent" and "hæmorrhagic" strictures will not require any further comment here.

SECOND LECTURE.

THE SURGICAL ANATOMY OF THE URETHRA — SPASM — THE DIMENSIONS OF THE URETHRA — OTIS'S VIEWS — THE CURVATURE OF THE URETHRA — THE RELATION OF THE URETHRA TO THE RECTUM — ATTACHMENTS OF FASCLÆ — OPENING OF THE SEMINAL DUCTS.

I SHALL ask your attention on this occasion to certain points respecting the structure, dimensions, and anatomical relations of the urethra. Some persons, it would appear, seem to think that anatomical knowledge affords but little help in this department of surgery. Now, I dare say, many of you learnt to pass a catheter with tolerable dexterity before you knew anything from dissection of the anatomy of the parts operated upon; and so, perhaps, you might continue to do, but the occasion will come when you will find yourselves entirely at a loss, and unable to give your patient that assistance which you might otherwise have done.

I could give you numerous illustrations in point, but this is unnecessary, as I take it to be the duty of every person who intends to practise surgery to make himself acquainted with the anatomy of the human body, without excepting any portion of it.

Nor is it my intention to give you an anatomical description of the urethra; I shall presume that you are familiar with the anatomy of this canal; if you are not, I shall refer you to the various treatises on the subject.

My object here is to point out in what directions your anatomical knowledge may be of service to you, and where you may expect to derive assistance from it in the practice of surgery as applied to diseases of the urinary organs.

The urethra is made up of various structures, possessing different properties, and arranged in layers— viz., internally mucous membrane, then involuntary muscular fibre, disposed longitudinally and circularly, and lastly, erectile tissue, which everywhere surrounds it. Further than this, it must be remembered that in its deeper portion the urethra is embedded in voluntary muscular fibres, which are capable of exercising a compressing force upon its walls, sufficient under the excitement of various stimuli, at all events, to prevent the expulsion of urine from the bladder. Such an impediment is known as spasmodic stricture. Uncomplicated with organic stricture, it is exceedingly rare. It is usually provoked by some active inflammation of the urethra, such as a gonorrhœa, or by the irritating influence of disordered urine; but when uncomplicated with organic stricture, it can never be regarded as affording a serious obstacle to the passage of an instrument into the bladder. Though this condition, when circumstances permit of it, may be successfully combated by rest and warmth and opiates, and such other measures as are calculated to remove muscular rigidity, where retention of urine exists, catheterism should be at once practised.*

In one instance that has come under my notice, in an eminent member of our profession, the rigidity of the muscles of the perinæum was so unusually great, and the distress on proceeding to pass a catheter so extreme, that I deemed it desirable to place the patient under chloroform. When this was done, a full-sized catheter passed readily into the bladder. On several subsequent occasions a similar course has been followed.

In reference to the muscular surroundings of the urethra, it should be remembered that it is the occurrence of spasm which determines retention. There can be no such thing as an impermeable urethra, except in connection with a urinary fistula. So long as the kidneys go on excreting urine, so long, unless a disturbance of muscular action takes place, will urine continue to find its way through the most contracted urethra.

---

* Guthrie enforces this so characteristically that I will repeat his own words. In relating such a case, he remarks:—"I was taught better many years ago by a Scotch friend, a young man, though an old soldier, who, after a debauch of this kind, which lasted half the night, found he could not make water when he awoke in the morning from his feverish dreams. He sent for me, begging I would bring a catheter with me. When I arrived I proposed an opiate; his answer was peremptory enough, "D— your draughts, doctor, pass the catheter; I have had it before." As remonstrance was useless, I passed the instrument and drew off his water, upon which he jumped into bed, saying, "God bless you, doctor, but d—— your physic." Since that time I have always made it a rule to try and pass a catheter in every case of retention of urine. If it passes, so much the better; if it does not, the patient submits more cheerfully to the longer course of treatment."—Guthrie, *On Diseases of the Bladder*, p. 89. 1834.

Every person who suffers from stricture finds out from experience the maximum quantity of urine over which he can successfully exercise propulsive power; and so long as this quantity is not exceeded, the ability to expel urine remains, although the stream may be exceedingly small, or even issue in drops. Should, however, from any cause, urine be allowed to collect in the bladder beyond the accustomed limit, the propulsive apparatus becomes disarranged by being called upon to do unaccustomed work, and irregular spasmodic efforts take the place of that combined muscular action necessary in the case of a person who at the best of times voids his urine under difficulties.* This consideration is offered as explaining how retention may be regarded as an accident occurring in the course of a stricture case, and how it is spasm becomes superadded to permanent urethral obstruction.

The view that organic stricture within the penile portion of the urethra is occasionally the excitant of spasm in the deeper portion, and that the latter is curable by the removing of the former by urethrotomy. has recently had some prominence given to it in a discussion between Professor Otis and Dr. Sands, of New York.† That a permanent obstruction in one

---

* "How frequently we see a spasmodic condition of the urethra supervening upon old organic stricture, and causing retention of urine. No doubt it depends upon irritation, beginning with the urethra behind the stricture, which exerts its influence, first upon the nerves of sensation, and thence upon the muscles of the urethra, through the excito-motor function of the spinal marrow. Large doses of opium relax this muscular spasm, and the patient is able to micturate."—Hilton, *On the Therapeutic Influence of Rest*, p. 250.

† *Hospital Gazette*, U. S. A., April, 1879.

part may disturb the muscular action of another is what might be expected in a system where the proper performance of function is dependent upon a normal condition and mutual relationship of the whole. I cannot, however, say that I ever met with a case where the spasm so produced was sufficient seriously to oppose the introduction of an instrument along the urethra after the organic penile stricture had been passed.

The mucous lining of the urethra has depressions in it which in the natural state cannot be regarded as affording any serious obstacle to the passage of one of the larger sized catheters. In cases of stricture, these lacunæ, behind the obstruction, become largely dilated, and are apt then to catch the end of the instrument after it has been passed through the stricture. Again, these dilated lacunæ may afford receptacles for urine, which, becoming decomposed, sets up inflammation within and around the urethra.

The erectile tissue, which everywhere surrounds the urethra, is not unfrequently the source of the hæmorrhage which follows where false passages have been made. The hæmorrhage under these circumstances rarely proves very considerable or persistent.

Sir Henry Thompson[*] furnishes us with the dimensions of the urethra taken by measurement at its various parts, from which it appears that the meatus is the smallest portion; "next is the point of junction between the membranous portion and the

---

[*] *On Stricture of the Urethra*, p. 6.

bulb, while the centre of the prostatic portion and the sinus of the bulb are the largest." The respective dimensions of the urethra are best understood by reference to casts, such as are pictured in Sir Everard Home's work on Stricture (Fig. 4).

After all, as Sir Henry Thompson goes on to state, "it is not the *actual* size of the various parts of the passage which is of the greatest consequence to the practical surgeon, and the foregoing measurements may be most advantageously viewed as possessing relative rather than absolute value."

Dr. F. N. Otis, of New York, in a recent work on the treatment of stricture, advances

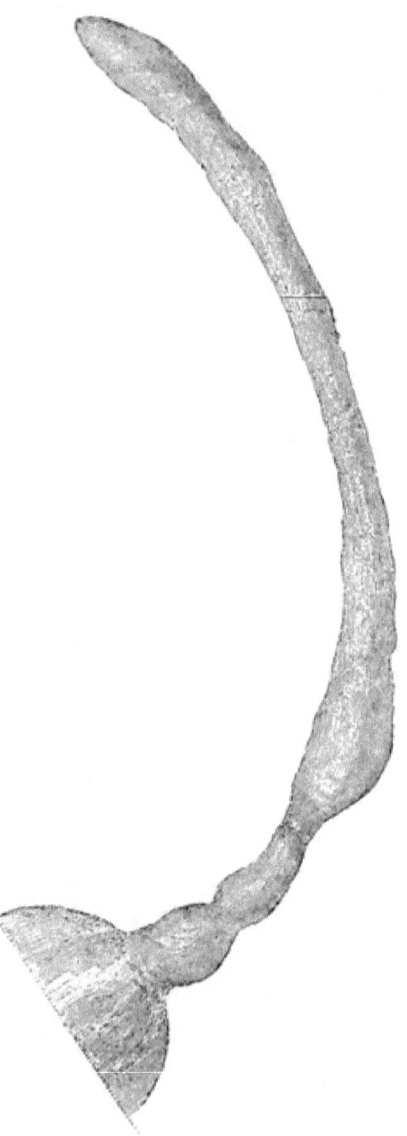

Fig. 4.

the proposition that every urethra has an *individuality*, and that no *average standard* is of use in examining a given urethra. This he demonstrates by the use of a very ingenious instrument, called a urethrometre, consisting of a straight tube, the end of which can be made into a kind of fenestrated sphere, the latter corresponding to a dial-plate at the handle, which marks the size of the sphere. By this means, the normal calibre of the urethra can be accurately measured, as also the circumference of the stricture.

To give an example: when the circumference of the penis was three inches, the calibre of the canal was found to be 30 m. of the French scale; when it was 3¼ inches, it would be 32 m., and so on in proportion.

Upon these views as to the measurement of the urethra Dr. Otis bases a method of treatment which will be referred to again when I come to treat of the various plans of practising urethrotomy.

If we refer to a side view of the pelvic organs, we can advantageously study the relations of the urethra to the surrounding parts (Fig. 5, from *Gray's Anatomy*). The curves of the urethra should here be noticed, with the view of determining the best position of the parts for the introduction of instruments along the canal. If the operator were to attempt to pass a catheter with the penis in a pendant position, it is quite evident that he would have to encounter two curves, a difficulty which in the case of a rigid instrument would be found insurmountable; whilst, on the other hand, the posi-

British Medical Journal, Feb. 26th, 1876.

tion of the penis may be so varied during the operation

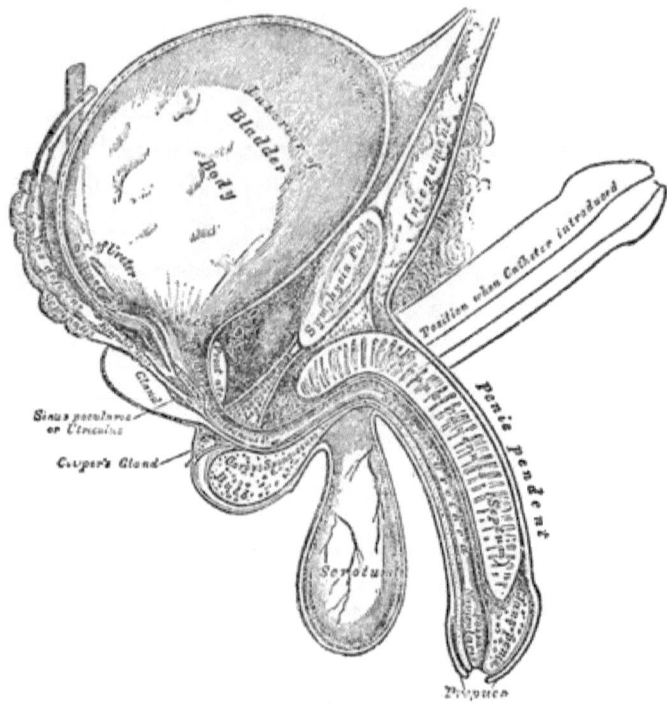

Fig. 5.

as entirely to do away with one curve, and materially diminish the resistance of the other.

It may be here observed that the urethra is only rigidly fixed at one part of its course—viz., as it passes through the triangular ligament. In front of this, it is sufficiently movable to permit of the whole of the canal being brought upon the same plane as the membranous portion which is contained between the two layers of the triangular ligament. Hence a straight instrument, such as the staff used by Key for lithotomy, may be introduced into the bladder quite as readily as an ordinary curved catheter.

It should be remembered that it is at the point where the urethra goes through the triangular ligament the inexperienced operator meets with his greatest difficulty in passing a catheter, the point of the instrument being usually allowed to drop so as to press against the ligament *below* the urethral aperture. After noticing its curvatures, we should next trace the urethra in its course to the bladder.

The first portion of the passage is subcutaneous, being situated along the under surface of the penis. Here it can be manipulated externally by the fingers, and direct assistance in this way rendered in the piloting of bougies through tight or tortuous strictures. In the case of strictures in this portion, their division by external or internal urethrotomy is accomplished without difficulty, by reason of the facility with which the urethra can here be handled. As the urethra passes behind the scrotum, it consequently becomes more deeply situated, and less accessible to external manipulation; whilst further again, in the perinæum, and behind the triangular ligament, little impression can be made upon it from without. If, however, you will refer again to the side view of the urethra, you will see that even here it is not beyond the reach and control of the finger.

On introducing a bougie into the bladder, and the index finger into the rectum, the line of the urethra for an inch and a half of its course can be distinctly made out, and the position of the instrument determined. Now, when we consider that the great majority of strictures are situated in this portion of the urethra, this is a piece of anatomical knowledge worth remembering. Fewer false passages would be made in difficult cases of catheterism if we bore in mind that we had the means of testing in the deep portion of the urethra the course the instrument was taking, and of rendering assistance to the passage of the instrument through the stricture by the introduction of the finger into the rectum.

To proceed. If the finger is carried still further, the line of the prostate can be distinctly made out, and any alteration in its size or consistence noted.

In difficult catheterism arising from prostatic enlargement, assistance may often be rendered in "tilting" the instrument into the bladder by the finger in the rectum. Further than this, where, by reason of the enlargement of the middle lobe, there is much resistance to the entrance of the catheter into the bladder, by placing the finger as a support to the prostate, that concussion or shaking of the gland is prevented which in two instances that came under my notice led to the occurrence of fatal pelvic cellulitis.

A reference to Figure 6 also shows that in retention of urine the posterior wall of the bladder can be explored and commanded sufficiently to admit of its being punctured without injury to the surrounding parts. In children, not only can a stone in the bladder be felt by the finger in the rectum, but, as Mr. Thomas Smith has pointed out, its removal may be facilitated.

Turning to the anterior aspect of the bladder, we should observe that the peritoneum, in its reflection from the back of the abdominal muscles on to the bladder, leaves a space just above the symphysis pubis, where the bladder, when it is distended, may be punctured without injury to the peritoneum.

We should now proceed to notice the attachments and connections of the various fasciæ having relation with the urethra; for we shall find that when matter forms around the urethra, or extravasation of urine occurs, the direction taken by these fluids is entirely influenced by the attachments of the fasciæ. If you refer to a side view of the pelvic fasciæ, you

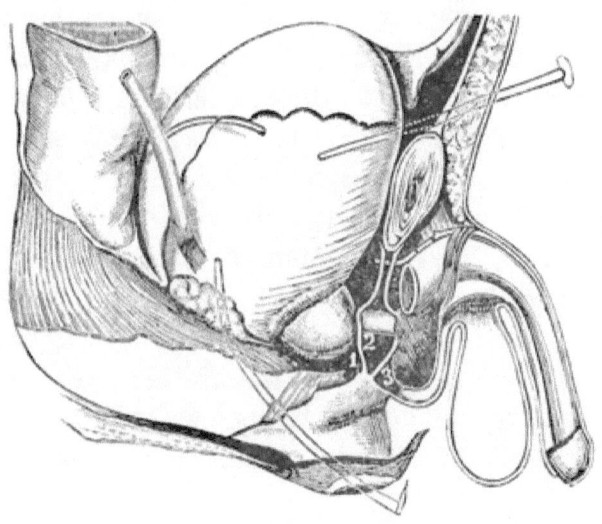

Fig. 6.

will see that there are three distinct compartments where, about the perinæum, fluid may collect. In each, the course taken by the fluid—be it urine or matter—will be different, the difference being determined by the connections of the layers of fasciæ between which it is placed. When matter forms around the prostate gland, in the compartment marked 1, it cannot come forwards, in consequence of the triangular ligament; it therefore goes backwards into the cellular tissue of the pelvis, where it will spread with great rapidity and fatality, under the name of pelvic cellulitis. The formation of matter in this position may be occasioned by injuries to the prostatic portion of the urethra. Not unfrequently it is the cause of death in lithotomy, where the incision into the prostate has been so free as to include its fibrous investment. When matter forms in compartment No. 2, between the layers of the triangular ligament, it either bursts into the urethra or makes its way towards the anus. Suppuration in this position not unfrequently takes place as a consequence of a very acute urethritis, and gives rise to the belief that inflammation of the prostate has occurred. Abscess here often simulates suppuration in the prostate, and is mistaken for it. Retention of urine, from pressure on the urethra, may in this way be caused, and relief is not obtained until either the abscess bursts into the urethra, or is opened by an incision.

When matter forms in compartment 3—that is to say, between the superficial perinæal fascia and the superficial layer of the triangular ligament—fluid is conducted towards the scrotum, and from thence it may pass to the anterior surface of the abdominal parietes, over a very considerable area. If unrelieved, the damage that is done to the tissues in contact with extravasated urine is immense; large portions of skin and cellular tissue slough and are discharged, and high constitutional symptoms, which tend rapidly to assume a typhoid character, not unfrequently are consequent upon the presence of urine in this position. Unless relief is speedily afforded by the knife, urine may travel as far as the umbilicus, or even above it, as we have seen on several occasions.

Passing to the interior of the urethra, I would remind you that within it are the openings of the canals conveying the semen, and other fluids which it is supposed are engaged in its elaboration. Hence we may infer that that which is an obstacle to the natural escape of urine is, *pro tanto*, an obstacle to the efficient discharge of the other secretion; and this undoubtedly we find to be the case, for sterility on the part of the male is constantly met with as one of the consequences of stricture.

There is one very important landmark which must not be passed by unnoticed. If you expose the perinæum, you will see that it is marked along the median line by a prominent ridge or elevation, called the raphé. This is a guide to us in many operations on this part. Along it the perinæum can be incised for stricture, abscess, or extravasation, to any necessary depth, with no risk of serious hæmorrhage occurring, whilst in the operation of lateral lithotomy it indicates the position of our first incision. When the perinæum is tumid from extravasation of urine or suppuration, it not unfrequently happens that the raphé is more or less pushed over to one side or the other, or even slightly curved. The incision to open the perinæum must correspond with such a deflection, otherwise troublesome hæmorrhage is likely to follow.

Such, then, are a few considerations which a knowledge of the anatomy and relations of the urethra suggests. They are sufficient to indicate the necessity of this study as the only proper preliminary to undertaking the management of its disorders.

THIRD LECTURE.

Symptoms of Stricture—Granular Urethritis—Consequences of Stricture on the Genito-Urinary Organs—Nervous and Spasmodic Affections Simulating Stricture.

The patient's suspicions that he is suffering from stricture are usually first aroused by his noticing some alteration in the force, direction, or size of the stream of urine in the act of micturition.

Unfortunately, however, for him, these indications do not generally become apparent until the disorder has made some considerable progress. It is of the first importance, then, that we should consider what may be regarded as the premonitory stage.

If we analyse the symptoms of stricture, excluding, for obvious reasons, cases of traumatic stricture and such-like causes of obstruction, we shall be able to refer them to one of two classes—viz., (a) symptoms indicating inflammation, and (b) symptoms of obstruction.

The symptoms of the first class—namely, those of inflammation—not only precede those of obstruction, but usually extend over a considerable period of time.

Most frequently they are consequent on an acute

gonorrhœa, and their extreme slightness, and the little inconvenience they occasion, are apt to render the patient almost unmindful of their presence. The only outward sign may be a continuous, though slight, muco-purulent discharge. Such a discharge is usually most obvious in the morning, and is often only sufficient to glue together the lips of the urethral orifice.

Further examination proves the existence, in varying degrees, of what are regarded as the cardinal symptoms of inflammation. Should we proceed to examine the urethra of such a patient with a bougie, there is *pain*, and an unpleasant sensation of *heat* as the instrument passes over the seat of the disorder; and that some inflammatory exudation—*swelling*—has taken place is evident by the resistance to the instrument that the operator is conscious of. That there is *redness*, or congestion of the part affected, is proved by endoscopic examination, should this be considered necessary for completing the diagnosis. These signs of inflammation—pain, heat, redness, and swelling—are usually accompanied by some obvious perversion of function. Not rarely there is frequency in micturition, as well as painful nocturnal emissions. These symptoms indicate the premonitory stage of stricture. The urethra becomes granular, a condition which has its counterpart in the granular lids of scrofulous children, with which we are so familiar.

The appearance presented by the healthy urethra, as viewed through the endoscope, "resembles very closely, on a smaller scale, a healthy rectum as seen through a speculum, or a vagina, but with the difference that the folds of mucous membrane of the healthy urethra are longitudinal instead of transverse."*

The granular urethra presents a somewhat blotched appearance, the radiating lines being concealed by points of granulation. The contrast between the healthy and granular urethra is seen in the following sketch (from Heath).

Fig. 7.

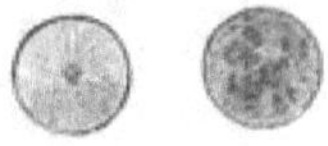

The inflammatory or premonitory symptoms, if unchecked, sooner or later give place to those indicating that obstruction has taken place; the stream of urine as it issues is wanting in force, or it is twisted or diminished in size, and as the contraction goes on these inconveniences increase.

In some cases the symptoms are of an anomalous character, and have no obvious correspondence with the description I have given. The formation of a hernia, or an enlargement of the testicle, may be the only indication of an urethral obstruction. It is also not uncommon to meet with cases where a sudden retention of urine is the first intimation to the patient that he is suffering from stricture.

Incontinence of urine, especially at night, is sometimes caused by stricture. I was recently consulted by a medical man in reference to this symptom, which I was able to explain and remove by the detection and treatment of a tight stricture he was not previously aware of.

It seems very strange that progressive changes in

* Heath, *On the Endoscope*, p. 7.

the urethra, such as stricture involves, should go on without the patient being aware of them, until, as it were, a crisis is reached. But so it is. A certain proportion of cases fail in affording a satisfactory explanation, further than to suggest that the patient has been wanting in the commonest observation.

The obstructive material, whilst these symptoms are continuing, gradually alters in its character. At first, it is merely an inflammatory exudation, soft and readily compressible; later on, as it becomes organised, it comes to resemble, not only in appearance, but also in its disposition to contract, the tissue of which scars are made up. The degree of resistance that this adventitious tissue is capable of offering is often remarkable, whilst in extent it is in some instances sufficient to convert the whole perinæum and scrotum into an indurated mass.

The worst feature connected with this deposit is its indisposition to become to any extent absorbed. You may exercise pressure upon it, you may divide it with the knife, or act upon it with caustics, but you cannot entirely remove it or deprive it of its inherent quality of contraction.

The treatment of advanced stricture will, therefore, be seen to resolve itself to a great extent into palliating and adapting, and this has been brought to such perfection that it is remarkable how little distress the patient may be conscious of, provided he exercise a moderate amount of care and precaution in the management of his own case. And, in speaking of treatment, it is well that we should understand our position in reference to the disorder we undertake to treat. Fortunately for mankind, there are many of its ailments which have a natural tendency towards what I may call a spontaneous cure, and even some of the most malignant diseases occasionally undergo changes of a benign character, remain quiescent, and cease to trouble. It is not so, as a rule, with stricture, unless it is kept in check by appropriate means; it is progressive, and the longer it remains untreated, the more hurtful it becomes.

It will be desirable here to note the remoter changes that take place.

As the contraction increases, the urethra behind the stricture becomes dilated, so much so that in some cases pouches are formed in which urine is apt to collect. In long-standing strictures these dilatations are so well marked as not to disappear after the stricture has been remedied. A gentleman, who some time previously had a tight stricture divided at the orifice of his urethra, consulted me in consequence of the very disagreeable odour of his urine when first passed in the morning. On examining him with a catheter, I found that urine collected in two very considerable pouches in his urethra. I advised him to drain his urethra every night, just before going to bed, by means of a catheter, slowly introduced and withdrawn. This had the desired effect, and speedily removed the inconvenience he complained of.

These pouch-like dilatations behind the stricture, which not unfrequently extend to the small ducts and lacunæ opening into the canal, are often the cause of peri-urethral, or, as it is otherwise called, urinary abscess.

The constant presence of urine in the urethra behind the stricture sets up inflammation not only within the canal, but also around it. Should suppuration occur, unless relief is given externally, the matter will find its way, by ulceration, into the urethra, and extravasation of urine follows. In the great majority of cases it is in this way, I believe, that extravasation happens, and not, as we are generally led to suppose, by rupture of the urethra behind the stricture.

It has been proposed to take advantage of the distended condition of the urethra behind the stricture to effect dilatation from behind forwards. Mr. Furneaux Jordan, of Birmingham, describes, in an interesting paper, this method of treatment, which he has practised with success.*

In addition to these changes in and around the urethra, the bladder becomes structurally altered, in consequence of its action and function being disturbed by the obstacle to micturition. In one case you will find its walls thickened and its cavity contracted, whilst in another it is expanded, with walls thinner than natural. In the former it is hypertrophied, for the purpose of overcoming the resistance offered by the stricture to the natural discharge of the urine; just as the heart, by an increase in its bulk, compensates for the resistance that is offered to it by an impeded circulation. Where the bladder is dilated and thinned, it seems to have gradually yielded to an obstacle which it has

* *British Medical Journal*, Nov. 9th, 1872.

been beyond its power to overcome. What determines the one or other condition is, I fear, little more than a matter of surmise.* Probably the degree of irritability the stricture occasions, or the disposition of the patient, has something to do with it.

It occasionally happens, though rarely, that rupture of the bladder takes place during the effort of a patient to overcome a stricture. An instance of this occurred in the Infirmary some months ago. Here the patient had been suffering from retention for some days. When admitted into the Infirmary he was in a state of collapse, from which he never rallied, dying eighteen hours afterwards. A catheter was introduced into the bladder without difficulty immediately after his admission, but only a few drops of blood-stained fluid escaped. At the *post mortem* examination, in the posterior wall of the bladder was found a rupture communicating with the cavity of the peritoneum. The edge of the opening was covered with lymph, and the rent measured, when not stretched, an inch and a half in length. There were also signs of peritonitis. Though there could be no doubt of the patient having suffered from retention, no sensible diminution of the calibre of the urethra could be discovered, so that we must conclude that the obstacle was occasioned by

---

* Mr. Cadge's explanation is a probable one:—" Muscular hypertrophy of the bladder, as of the heart, may exist under two conditions, simple hypertrophy or hypertrophy with dilatation. The former is seen most frequently in cases where the obstruction is due to stricture, where the patients are of middle age, and in the possession of fair vital power; the latter in the aged, whose enlarged prostate is the impediment, whose vital power is on the wane, and in whom the bladder is apt to exhibit atony or even paralysis."—*British Medical Journal*, October 2, 1875.

spasm. The history of the case would admit of no other conclusion. The specimen is preserved in the Museum of the School.

Going still further back, we find the ureters and kidneys yielding to the pressure of the *vis a fronte ;* these may, in the course of time, become mere tortuous tubes, and little else than subsidiary bladders, as you may gather from the specimens I am placing before you.

Kidney disease, varying from slight congestion to almost complete disorganisation, is a frequent concomitant of stricture ; this is a fact which should never be lost sight of, and brings me to notice the importance of examining the urine.*

No operation, however slight, can be regarded as absolutely free from danger, and even the passing of a catheter, simple as it seems, forms no exception to this rule. In a very interesting article, to which I shall have occasion again to refer, on Urethral Fever,† my colleague, Mr. Banks, narrates a case, I well remember, where death occurred six and a half hours subsequently to the passage of a bougie. Allusion is also made in the same paper to a case of Mr.

---

* "Among the cases of stricture, one hundred deaths occurred (in Guy's Hospital) in nineteen years, giving a yearly average of about 5·26. Of the whole number of cases, the kidneys were suppurating in forty-one of the hundred ; they were wasted away or inflamed in eighteen ; in seven they showed evidence of the changes included under the term Bright's disease, or were cystic ; while, in the remaining thirty-four, they were healthy. Thus fifty-nine, or nearly three-fifths of all the cases, had advanced disease of the kidneys."—J. F. Goodhart, *Guy's Hospital Reports*, Series iii., vol. xix.

† *Edinburgh Medical Journal*, June, 1871.

Padley's, in the Infirmary, where death followed, under similar circumstances, in the course of a few minutes. These undoubtedly were cases of shock, so severely felt as to be almost immediately fatal, and before the development of those reactionary indications on the system generally which have suggested the term "Urethral Fever."

As no subjects bear shock so badly as those who are suffering from structural kidney disorder, every means should be taken for determining the precise condition of these organs before deciding, in a case of stricture, the line of treatment that is to be pursued, and hence a thorough examination of the urine is of the first importance.

The existence of advanced kidney disease will limit us to such proceedings as have for their object the preservation of life, independently of other considerations, and under such circumstances the range of treatment becomes restricted. I shall in another lecture make observations on the examination of the urine, and, as occasion arises, comment upon certain points connected with the treatment of the various abnormalities that may here be discovered.

Before concluding my remarks on the symptoms of organic stricture, I must refer to certain nervous and spasmodic affections of the urethra which simulate it. Of this class of cases, which are sometimes misleading, you see comparatively few in hospital practice, and it is from other sources that I must chiefly draw the illustrations which will be necessary for my purpose. Because these cases are wanting in certain physical signs, do not for a moment suppose that those who complain of them do so for the purpose alone of exciting sympathy or commiseration; on the contrary, I can assure you the distress that is thus occasioned often renders the lives of those who suffer from such symptoms very miserable, and consequently they are deserving of your most careful consideration. I do not know how it is, but when one comes to speak of disorders which are described as of "nervous" origin, we almost unconsciously add the caution that I have given, as if under the generic term of "nervous affection" the most horrible and distressing sensations are not included. In our search for the material we are apt to forget the immaterial. Who has not seen patients annoyed beyond measure by that condition which has been so graphically described by Sir James Paget as "stammering with the urinary organs."* One such stammerer, an elderly gentleman, without a large prostate, I attended for a long time. He used to spend his days in selecting words, the repetition of which, he thought, during micturition, favoured the act. He kept a list of these words by his side, crossing one off when it appeared to him to have lost its effect, or was supplanted by one more potent. Many words were coined for this purpose. It was impossible to refrain from a smile on seeing, as I had often to do, micturition being performed whilst a word was rapidly repeated as if invoking the assistance of some ancient Deity who was specially interested in such matters. And yet to the end of his life my old and valued friend

* *Clinical Lectures*, by Sir James Paget, Bart.

believed in the efficacy of this proceeding. Another gentleman I attended was in the habit of provoking micturition by the sound of water falling into a basin; whilst a lady, after an operation where catheterism was necessary, first spontaneously passed water after the nurse had given her a mouthful of cold water; for many days afterwards she could never repeat the act until she had previously taken a drink. I could multiply such examples, but it would be at the expense of your time, if not, you might think, of your credulity; but I have said sufficient to show the importance of recognising the morbid effect of nervous influence on micturition. You will not cure your patients by asserting your belief in the non-reality of their ailments, but having carefully established the absence of any organic disease, you may, by your counsel and advice, as well as by your medicine, often do much in restoring the nerve-tone which is deficient, and so prevent the patient drifting into a hopelessly depressed mental condition.

Again, there are a variety of morbid influences, chiefly constitutional, which undoubtedly interfere with micturition by inducing spasm, more or less persistent, in the muscles connected with the deeper portion of the urethra and the neck of the bladder. Under the title of *Contracture du Col Vésical*, various morbid conditions have been described, chiefly by French authorities, as to the existence of which no possible doubt can be entertained, however you may choose to explain them. Of the more recent treatises on the subject, that by Dr. Delefosse may be advantageously studied by those requiring further information.* Examine these cases as you will, with all the light the literature of the subject can throw upon them, you can come to no other conclusion than that the word "spasm," with which we are so well acquainted, as accurately as anything else, describes the nature of the impediment to micturition which is their prominent feature. The exciting causes of such spasm are local and constitutional, the former including irritations proceeding from the rectum, such as piles or fissures, whilst amongst constitutional excitants the taints of gout and of rheumatism are the most frequent. The French authorities to whom I have referred lay great stress on rheumatism as an exciting cause of this affection.

In addition to those general principles of treatment which rheumatism and gout require, very great advantage will be found in these cases from a course of treatment at certain watering places. Of these I may mention as being best adapted to this purpose, Vichy, Contrexéville, Vals, and Evian.

As to local treatment, I have little to say. When the exciting cause of the spasm has been removed, I have found benefit from the application to the deep portion of the urethra, for the purpose of removing the extreme sensitiveness which sometimes remains, of a solution of nitrate of silver (five grains to the ounce).

In arriving at the conclusion that a case of impeded or hesitating micturition is functional and independent

* *Sur la Contracture du Col Vésical*, par le Dr. Delefosse. Paris, 1879.

of structural alteration, you must remove all source of error by a careful physical examination of the urethra. If you learn to acquire delicacy and tact in the use of urethral instruments necessary for the purpose, you will never give your patient cause to regret that he has submitted to such an exploration; on the contrary, in the absence of true stricture, he will be comforted by the assurance that he can be completely and permanently relieved.

FOURTH LECTURE.

Examination of the Urine.

An examination of the urine is an essential preliminary to the investigation of all diseases affecting any portion of the urinary tract. Such examination may at once reveal the cause of the disease, and if it fail to do this, it is no less valuable in determining the particular line of treatment to be adopted. For instance, in a case of stricture, or of stone, a normal or an abnormal condition of the kidneys will indicate respectively the means we shall select for removing the cause of the patient's symptoms. Save when there is some urgency or necessity for prompt interference, I never submit the urethra or bladder to any treatment requiring the introduction of instruments until I have first ascertained the state of the urine, until I have formed an estimate, not of the condition of the part that may alone be the seat of the disease, but, if I may use the term, of the general health of the whole urinary apparatus. The urethra or the bladder, as the case may be, is only a part of a system, and a knowledge of the whole is necessary to the proper understanding, pathologically and therapeutically, of any particular part. Those who are accustomed to see my practice know the importance I attach to this observation, and frequently see the good reasons that show themselves for acting up to it.

It would be impossible for me, in a course such as this, to compass all that you should know of the physiology, chemistry, and pathology of the urine. I shall only attempt to place before you certain broad clinical features having reference to abnormal conditions of the urine, with some practical deductions obtainable therefrom.

I will take for a starting point a definition of what normal urine should be ; and I cannot do better than select the words of Dr. Roberts :—" Healthy urine is a clear, watery, amber-coloured, saline solution, generally acid, with a specific gravity of about 1020."* Urine differing from this description cannot be regarded as typical of that which is healthy, though the deviation may be so slight or temporary as not to be accompanied by other signs of ill-health.

First, with regard to the *specific gravity* of the urine. This is taken by means of the urinometer, and indicates the density of the urine, from which a rough estimate of the solids it contains may be made. The last two figures of the specific gravity being doubled, the quotient approximately represents the amount of solid matter per 1000. Thus urine of specific gravity 1020 would contain about forty grains of solids. Specific gravity must be considered as indicating only the density of the urine, regardless of the relative proportion of the constituent salts or other solids

* *On Urinary and Renal Diseases*, by Dr. W. Roberts.

it may contain; for, as Dr. Carter remarks, "two urines having the same specific gravity may yet differ very much from each other in the relative amount of their solids, one having a large quantity of organic products and a relatively smaller quantity of salts, and vice versâ."*

When a person is in the habit of passing urine of a low specific gravity, it is necessary to be especially cautious in his case to avoid doing anything which would shock him, and so interfere with the feeble excretory action of his kidneys. Such persons as these are often on the borderland of uræmia, or urinary poisoning. Sir James Paget very pertinently remarks, "Let me tell you of a symptom which must make you specially cautious if you have to catheterise elderly or old men. If they are passing large quantities of pale urine, of low specific gravity, whether containing a trace of albumen or not, they will be in danger from even the most gentle catheterism."†

When I have, as we all have, to pass an instrument for such persons, the condition referred to makes me additionally careful to provide against any ill after-consequences, by adopting those means which we know of as being likely to ward them off. And in this I am generally successful.

In hysterical persons the low specific gravity of the urine is, I believe, the most frequent cause of the irritability of the bladder from which they suffer. The nearer urine approaches in every respect to a

---

* *On Renal and Urinary Diseases*, by Dr. W. Carter.
† *Clinical Lectures*, by Sir James Paget, Bart.

healthy standard, the less irritating it is. A patient who has to wash out his nasal passages for ozæna, finds a saline solution far less irritating for the purpose than pure water, and so it is with the urine.

When urine is irritating from being too concentrated, we bring it down to its normal standard by diluents such as barley water or linseed tea, with much comfort to the patient.

Having noted the specific gravity, the *reaction* of the urine is then determined by test-paper. It may be stated generally that healthy urine is acid, being rendered so by the presence of free acids and acid salts. The alkalinity of urine is traceable to two different causes: (1) to the presence of a fixed alkali, a condition usually associated with a debilitated state of health; and (2) to the presence of a volatile alkali (ammonia), the result, with hardly an exception, of the decomposition of urea.

The mode in which the urine becomes ammoniacal from decomposition is easily explained. "One atom of urea with two atoms of water, by a simple re-arrangement of their particles, become converted into two atoms of carbonate of ammonia; 1 atom urea $C_2 H_4 N_2 O_2 + 2 HO = 2 (NH_3 CO_2)$. This change is so easily brought about that mere boiling of a solution of urea in distilled water is sufficient to effect it." *

"To distinguish between the volatile and fixed alkali, the test-paper, after being rendered blue, should be allowed to dry in the open air. If the blue colour

---

* Dr. W. Roberts's *Op. Cit.*

persist after complete desiccation, the alkali is fixed; if it disappear, and the original colour be restored, the alkalescence is due to ammonia. The smell of the urine is also a useful indication in such cases."*

Persons passing urine which is alkaline by reason of the presence of a fixed base, usually require medical rather than surgical aid. Generally, alkaline urine indicates a weakly physical condition, combined, it may be, with a highly strung state of the nervous system. It is in the latter subjects that opium is of the greatest value. By counteracting the injurious effects of nerve tension it often restores a normal reaction to the urine, and in this way is realised Sir Thomas Watson's observation, that "no single drug probably has so much power in rendering alkaline urine acid as opium." †

As bearing upon what has been said about the decomposition of urea as a cause of alkaline urine, I may mention a case I saw some little time ago, which much interested me. It was one of stricture, with extravasation of urine into the scrotum, occurring in a person suffering from Bright's disease of the kidneys. Though the extravasation had come on suddenly, and had existed for twenty-four hours unrelieved, there were no signs of acute inflammatory action and commencing gangrene, such as are usually expected. However, the tension being considerable, I incised the parts involved in the extravasation. As the fluid escaped from the incisions, I noticed that it

---

* Dr. W. Roberts's *Op. Cit.*
† *Principles and Practice of Medicine*, 5th ed., vol. ii., p. 712.

had not that strong ammoniacal odour which is so perceptible in such cases. Subsequently I treated the stricture, which was exceedingly tight, and for some time kept in abeyance the more threatening urinary symptoms. I was somewhat puzzled for an explanation, as I felt sure that the case was one of extravasation, and not of acute scrotal œdema. How was it then that extravasated urine failed to create gangrene? I collected some of the urine as it trickled through the wound, and compared it with some subsequently drawn off by the catheter. I found them identical, and in both there was an almost complete absence of urea. This then, to my mind, solved the mystery, and explained that as there was no urea to decompose there was no source for the production of the ammonia by which the destruction of tissues in connection with extravasated normal urine is effected. By the absence of urea the urine was rendered chemically harmless to the tissues with which it came in contact.

The urine, when rendered alkaline by a volatile alkali is probably to the surgeon the more interesting condition, as it is met with in those cases where this excretion has been retained within the bladder until it has decomposed in the way I have mentioned. Surgically we meet with this in fractures of the spine, in advanced stricture, and prostatic disease, where there is a mechanical obstacle to the bladder being emptied. In a case shown at the Liverpool Medical Institution, during the session of 1878–79, by Dr. Spratly, of Rock Ferry, where a fractured spine had been advantageously treated by extension and a Sayre's jacket, not only was the ammoniacal condition of the urine, due to paralysis of the bladder, a source of persistent trouble, but this state, in spite of constant care by catheterism and irrigation, led to the formation of numerous phosphatic calculi with marvellous rapidity, which caused the patient much annoyance.

Further than this, we find the presence of ammonia in the urine acts as an irritant to the structures in contact with it, and inflammation of the coats of the bladder, more or less acute, adds to the patient's distress. A moment's reflection shows how much may be done, not only to prevent but to remedy such a serious complication to the states of disease I have mentioned, by the employment of catheterism and ablution of the bladder and urethra. The sense of smell alone is often sufficient to indicate to the practitioner the necessity for adopting these precautions.

Therapeutically, the recognition of the cause of the abnormal reaction of the urine, whether it be due to a fixed or a volatile alkali, is a matter of much importance. The fixed alkali, as I have already remarked, is for the most part met with in persons with feeble powers of life, persons who invariably improve in condition with a generous diet and tonics containing one or other of the mineral acids. To give acids with the view of correcting both the health and the urine of the patient who always has in his bladder a residuum of urine, necessarily of an alkaline reaction, due to decomposition, is worse than useless; yet I have seen this done and justified on the ground that the urine was alkaline, no thought being attached to the consideration that the cause of the alkalinity was essentially one that could only be removed by mechanical means.

Though urine that is acid can be rendered alkaline by the administration of alkalies, the converse does not hold good except in a sense which is not generally appreciated. That urine is often rendered exceedingly acid and irritating by drugs which have the effect of disturbing digestion, I have no doubt. If I remember correctly, it was Dr. Bence Jones who showed that it was in this way alone that urine could be rendered acid. In practice I have particularly noticed this ill-effect in the administration of specific medicines, such as copaiba and cubebs, for gonorrhœa. These drugs are exceedingly nauseous, and their use is not unfrequently followed by disturbed digestion and a highly acid condition of the urine, which prejudices their healing effect on the urinary passages. For some time I was at a loss to explain how the old-fashioned, but very disagreeable, copaiba mixture, which contained sufficient liquor potassæ to render the balsam miscible, was so much more efficacious than the more elegant and modern forms of capsules and confections. The efficacy of the old mixture is, I believe, largely due to the alkali it contains counteracting any excess of acid which the drug creates. In practice I have proved this to be the case, not only with copaiba but with other drugs of a like nature and purpose. In giving cubebs, I find that its combination with bicarbonate of potash greatly increases its efficacy as an anti-blennorrhagic.

Ammoniacal urine has a further chemical importance, as in this state of alkalinity the triple phosphates are thrown down, and the formation of this variety of calculus is favoured. Hence persons who are passing urine that is ammoniacal are not only liable to attacks of cystitis, from the ammonia that is evolved, but, in addition, to the formation of phosphatic stones.

Proceeding further, *albumen* and *sugar* should next be searched for, the probability of finding the one or the other being suggested by the specific gravity, urine containing albumen being usually low, whilst when sugar is present it is the reverse. This, however, must be accepted with the caution that it is only generally true.

The persistent presence of albumen, when not due to the existence of some other fluid in the urine, such as pus, blood, or semen, points to a mechanical impediment to the circulation through, or disease of, the kidneys, and therefore is a symptom of grave significance. The detection of albumen under these circumstances must be followed up by a careful microscopical examination of the urine, in search of casts or other direct evidence of renal mischief. For reasons that I have already urged, the suspicion that kidney mischief may exist, as complicating some surgical disorder, must be carefully weighed, in order that the surgeon may be able to judge what, if any, operative measures are to be adopted in a case where, under other circumstances, such would have been desirable.

It should be remembered, in connection with the

subject of albuminuria, that some recent observations, conducted on a sufficiently large scale for the purposes of life insurance, and upon army recruits, have shown that a certain proportion of otherwise healthy persons, and with no other signs of structural kidney disease, have albumen in appreciable quantities in their urine.* From a careful perusal of these statements, I cannot suppose that they are in any way open to fallacy; we must, however, always attach considerable importance to the existence of albumen in the urine, however unexplainable it may be.

That there is *sugar* in the urine is rendered probable when the specific gravity is over 1030. In reference to its presence I shall not offer any further observations. In the treatment of such affections of the urinary organs as have come under my notice, I have rarely met with it as a complication, and when present it appeared in no way to influence the surgical treatment the local complaint required. In testing for sugar, the cupric test pellets, introduced by Dr. Pavy, will be found exceedingly convenient.†

I shall now pass on to enumerate the various deposits. These, when they exist in any considerable degree, are often apparent to the naked eye, and are a source of apprehension to the patient, who cannot fail to recognise in most instances the abnormality, though he may be unable to appreciate its significance. The deposits will be considered as (1) unorganised, and (2)

---

* *Medical Times and Gazette*, June 21, 1879.

† *Proceedings of Clinical Society*, Jan. 23, 1880. These pellets may be obtained from Mr. Cooper, Chemist, 26, Oxford Street, London.

organised. Not much difficulty, after a little practice, will be experienced in making out the nature of these several deposits; it is in referring them to the precise cause producing them that some hesitation will occasionally be felt. Until you have localised the source of a morbid deposit, your treatment must be empirical. I have known a patient go on having his bladder washed out and treated as if for a cystitis when the pus was derived from a chronic pyelitis. Hence, in examining urinary deposits, we must be specially careful in removing as far as possible all sources of fallacy. If, as Sir Henry Thompson remarks, we want the urine as contained in the bladder, we must take care that we get it. Furthermore, in the collection and examination of urine the most scrupulous cleanliness is to be observed, otherwise we shall introduce various sources of error which I might mention.

Of the unorganised deposits—namely, *uric acid, urates, phosphates,* and *oxalates*—I shall not occupy time by any extended notice; to do so would necessitate a reference to their chemistry and physiology beyond the limits of a practical course. I shall, in fact, not ask you to do more than, from the specimens placed before you and the drawings that accompany them, learn to distinguish their respective appearances. Though the variations in the quantities of these salts in the urine—their excess or diminution—chiefly come under the notice of the physician in connexion with a number of disordered actions, yet in their relation to the formation of stone, and the consequences produced by obstructed micturition and decomposition of urine, the surgeon must be largely interested.

There is a specimen of cystine to which I would specially ask your attention, as it is rare, and will be again referred to when I narrate the case of the patient from whom it was removed.

I will take the organised deposits in the following order:—

Mucus and epithelium, blood, pus, spermatic fluid; these deposits are all of such importance in the investigation of surgical affections of the urinary organs that I shall make some observations on each.

Mucus and epithelium.—If you allow a specimen of healthy urine to stand a short time in a glass, a thin flocculent deposit will be observed falling to the bottom of the vessel. This is mucus from off the urinary passages. It is readily distinguishable from all other deposits, the characteristics of which will be noticed in their place. In a doubtful case it must be remembered that urine containing pus yields albumen to the appropriate test, whereas if the cloudiness is due to mucus alone, it does not do so. An excess of mucus in the urine points to some source of irritation which should be determined if it continues. I have known an increased quantity of mucus in the urine to be the only symptom of stone. You would hardly think it possible that a person could be subjected to a prolonged course of deception by the normal appearance of his own excretion, but I have just now been consulted in reference to a case where a very large sum of money has been obtained from a young gentleman by a notorious quack pointing to the cloud of natural mucus deposited in the urine as being evidence that he was suffering from spermatorrhœa. Of course this impostor will escape unpunished.

The importance of making observations in regard to changes in the quantity and quality of the urinary mucus will be more fully referred to in a subsequent lecture on the formation of calculi. Hitherto I believe sufficient attention has not been given to this deposit in relation to this subject.

Blood.—The presence of blood in the urine is very naturally a source of anxiety to the patient, who usually at once detects its presence by the discoloration it gives rise to; the circumstances under which it appears in the urine must be very closely investigated. In examining blood in the urine, bear in mind the sage piece of advice given by the late Mr. Hilton: " Swim out in water all clots whose origin is doubtful, in order that you may see the shape. Over and over again you will find yourself able to diagnose the case by this simple common-sense expedient."[*] Blood that has been clotted in the ureters is sometimes seen in the form of worm-like casts of these tubes.

In falls or blows, where the kidney is lacerated, there is not uncommonly blood in the urine, and when by means of the microscope casts are discovered of the mouldings of the uriniferous tubes, the most conclusive evidence as to the nature of the injury is thus afforded. When blood comes from the kidney it is usually so intimately mixed with the urine as to give the latter a

---

[*] *Guy's Hospital Reports*, 1868, p. 20.

smoky appearance. When from the bladder or prostate, as in stone or tumours, blood generally follows micturition, reversing this order of proceeding when the source of hæmorrhage is the urethra, as in acute gonorrhœa. Hæmaturia must not be confounded with hæmatinuria, usually a paroxysmal affection, where the urine is discoloured, not with blood-corpuscles, but with their contents — namely, hæmatin.* The most valuable hæmostatics, so far as my experience goes, are ergot of rye, matico, and tincture of iron. The preparation popularly known under the name of "Ruspini's Styptic," I have also found of service.

In one of the worst cases of hæmaturia, from the kidney, associated with stricture I ever saw, the most marked benefit, when matters were beginning to look very serious, followed the administration of the infusion of matico — a wineglassful given every two hours.

Pus is seen in the urine in two different conditions; in acid urine, as minute particles which fall to the bottom of the vessel if the urine is allowed to stand, and giving it a somewhat milky appearance on agitation. Pus in alkaline urine forms a thick gummy mass, which adheres tenaciously to the bottom of the vessel. This difference in appearance in the two urines suggests a test by which the presence of pus is readily recognised. Urine containing pus becomes more viscid and tenacious by the addition of about half its quantity of liquor potassæ, whereas urine con-

---

* An interesting paper on this subject, with numerous references, by Dr. Stephen Mackenzie, will be found in the *Lancet*, July 26, 1879.

taining mucus is rendered less viscid-looking by the alkali. By the microscope we recognise the presence of the pus-corpuscles. As urine containing pus rapidly becomes alkaline by decomposition, its reaction should be taken immediately after it is passed. The reaction of purulent urine is by no means unimportant, as suggesting the probable source of the matter. Pus in acid urine most probably comes either from the urethra or the kidney. If from the former, a drop of matter can generally be made to exude on pressure from the meatus. Pus in urine that is alkaline immediately after it is passed is most probably from the bladder, being frequently met with in stricture, enlargement of the prostate, and tumours of the bladder. The source of the discharge in all cases of purulent urine must be thoroughly inquired into. As I have already said, I have seen a bladder subjected to active local and general treatment when the presence of pus in the bladder was due to a pyelitis in a largely dilated kidney.

Spermatic fluid.—Lastly, albumen may be traceable in urine to the spermatic fluid. Persons frequently imagine that they are passing considerable quantities of seminal fluid in the urine, and consequently you should be able to recognise it when present. In the great majority of instances you will be able to assure your patients that their alarm is groundless, and that they must look in another direction for an explanation of any symptoms affecting the genital organs of which they may complain. Great and sometimes irreparable damage has been done, not only to individuals, but to society at large, by unprincipled persons, whom the law ought to be strong enough to restrain, preying upon that instinct with which we have been the most strongly endowed. The word spermatorrhœa has done service in debauching both the body and the mind, to an extent which it would be difficult to realise, and in doing this the more effectually, the basest frauds and counterfeits have been resorted to. I allude to this matter, inasmuch as not only should you be able to detect spermatorrhœa when it really does exist, but in order that you may be enabled to give assurance to those who have been the victims of the deceit to which I have alluded.

To examine for spermatozoa, the urine should be allowed to stand for a time, so as to permit of the glairy seminal fluid falling to the bottom of the glass. The lowest stratum of the urine may then be submitted to microscopical examination, when, should spermatozoa be present, there will be no difficulty in recognising them.

I have been asked whether, in a very prolonged and aggravated case of true spermatorrhœa, it might be impossible to furnish this proof, by reason of the inability of the exhausted sexual apparatus to furnish the essential and characteristic element of natural semen. To this, with good reason, I would answer, it might be possible. But under such circumstances as these I should expect to find some corresponding sympathy in other portions of the generative apparatus. And such evidence the great majority of cases of alleged spermatorrhœa fail to afford. I do not purpose entering upon the subject of treatment;* it is sufficient for me to indicate how an examination of the urine, carefully conducted, may enable you to distinguish between the true and the assumed disorder. Let me give you one caution. If, by your examination of all symptoms, and of such evidence as the urine will afford, you come to the conclusion that your patient has no grounds for believing that he is voiding unnaturally his spermatic fluid, do not abruptly accuse him of wilful self-deception. Remember that mentally, if not physically, his condition is a morbid one, and that he requires just as much counsel and sympathy as if he were suffering from any other affection. If a person is weighed down by the dread of some imaginary disorder, you will not gain his confidence by telling him, without explanation, that he is deceiving himself. You will, by so doing, rather drive him into the hands of those by whom his fancies will be encouraged and his alarms intensified. He requires re-assurance, and this you must give him, not by dissembling, but by such a rational explanation of his condition as an educated medical man is capable of affording.

You will occasionally meet with very exceptional conditions of urine, both as to odour and appearance, which you must not be unprepared for. Some things are added to it by the patients themselves, for the purpose of fraud. When such is suspected, I need

* In the treatment of nocturnal emissions I have found the greatest benefit from douching the lower part of the spine, before going to bed, with water of the temperature of 120° Fahr.; when this is effectually done, the relief of this symptom is often immediate.

hardly say the mystery can be cleared up by causing the patient to pass water when some one is present.

Certain articles of diet and medicines produce an appreciable change in the urine; of these I may mention in illustration asparagus and copaiba. Whenever a patient complains of his urine smelling offensive and persistently so, be sure, before giving an opinion, that the urine is not abnormally retained in any portion of the urinary passages by an impediment to micturition, and so be allowed to decompose.

Fæcal matter.—I would remind you that particles of fæcal matter occasionally make their way into the bladder from the bowels, by routes resulting from abscess or ulceration, and discolour the urine. One such case I have recently seen, in an otherwise healthy person, with Mr. Rushton Parker, where the urine furnished undoubted proof of the existence of such a communication. The colour and odour, and detection by the microscope of particles of food, will furnish indications as to the nature of this deposit. In another case my attention was first called to this condition by the patient asserting that he occasionally passed wind into the bladder, which was followed by most severe colicky pains. A fæcal fistula need not necessarily communicate directly with the bladder; in a case recorded by Dr. Ord, the communication was between the bowel and the ureter.[*] In women such cases are more common, in consequence of the damage sometimes resulting from parturition.

I shall not occupy your time by a reference to the

[*] *British Medical Journal*, Sept. 7, 1878.

miscellaneous objects you may possibly meet with in the urine; examine them carefully, and you will generally be able to determine what they are, and having done this, you may often form a very good inference as to how they got there.

Apropos of this, I cannot help repeating a story I dare say you have heard before. A house-surgeon was examining, under the microscope, for a practitioner of the old school, a specimen of urine. Turning the slide about under the field, he casually remarked to the old gentleman, "Ah! I see you removed this with the catheter." The practitioner was so astonished at this observation that, rather than betray any ignorance, he went away puzzled and silent, forgetting that his young friend, recognising by the microscope the oil-globules, had drawn an almost irresistible conclusion which was correct. The practitioner, regarding this circumstance as showing good powers of observation and deduction, soon after made overtures to the house-surgeon, and a most profitable partnership was the result of this casual but correct remark.

I have only endeavoured to give a very cursory glance at the examination of the urine; for fuller and more systematic information I must refer you, for a ward manual, to the small work by Dr. Wickham Legg,[*] and to the able and classical text-book of my friend, Dr. Roberts.[†]

---

[*] *A Guide to the Examination of the Urine*, by Dr. Wickham Legg.
[†] *On Urinary and Renal Diseases*, by Dr. William Roberts, F.R.S.

FIFTH LECTURE.

Treatment of Stricture — Gradual Dilatation — Instruments Employed — The Filiform Bougie — Anæsthetics — Continuous Dilatation.

I now come to consider the treatment of stricture, and as dilatation by bougies is the oldest and most extensively employed means, and as all other methods of treatment are more or less subservient to this, it will be proper to give it the first consideration.

Before, however, submitting a case for instrumental treatment, the very important question should be asked, Is the patient in a suitable condition for its advantageous employment? With few exceptions, persons suffering from stricture seek professional assistance at times and under circumstances when they are least fitted to undergo the treatment necessary for their cure.

The patient who sends for you to relieve his retention has, most probably, induced this state by an excess of some sort. The urethra has been rendered irritable by the passage, perhaps, of unhealthy urine, which has provoked sufficient spasm of the urethra to convert impeded micturition into complete retention.

I have frequently pointed out the great advantage following the employment of rest, and such-like measures, in cases coming to the Infirmary for the treatment of retention.

There was admitted the other day, for retention, a patient on whom prolonged efforts had previously been made to pass a catheter. I saw from his clothing that he had been bleeding profusely. His bladder could be felt distended above the pubes, though not largely so. Under these circumstances, I ordered him a hot bath and a dose of laudanum, and afterwards to be well covered up with hot blankets. Thus he was enabled gradually to empty his bladder, and on the third day after his admission, at the first trial, I passed a small bougie through a tight stricture without drawing a drop of blood. Prolonged catheterism is in itself an evil; every deviation the instrument makes from the course of the urethra occasions a rent, and every rent leaves a scar, so that in this way the original stricture may be considerably increased.

When circumstances will permit of it, the employment of rest, and attention to the condition of the urine and the general health, not only facilitate the passage of instruments along the urethra, but render their use much more serviceable.

Under ordinary circumstances, I find that the recumbent position is the best for the patient to be placed in for catheterism. He is more at his ease, and more complete muscular relaxation is thus obtained. In cases of enlarged prostate the erect position sometimes adds to the difficulty of introducing a catheter by reason of the third lobe, when this is enlarged, pressing forwards over the end of the instrument. On placing the patient in the horizontal position, the lobe falls back and permits the catheter to pass with ease. Occasionally it is necessary to place the patient under the influence of an anæsthetic. As this cannot safely or conveniently be done in any other than the horizontal posture, it is as well you should accustom yourself to this position of your patient, otherwise you may expect to find yourself somewhat awkward in your manipulations.

I need not describe to you how to pass a catheter; observation and a little patience on your part will enable you to overcome those slight impediments which even the normal urethra presents. John Bell, in his *Principles of Surgery*, very aptly remarks, "There is no operation with which I should more earnestly entreat the young surgeon to make himself acquainted than this of introducing the catheter." *Festina lente*— be patient and never resort to force. Endeavour to attain your object without giving pain or causing hæmorrhage. There is no necessity for either; or rather I would say, the surgeon should know from the first feel of the parts with his catheter whether there is likely to be pain, and if so, there is just as much need for an anæsthetic in this procedure as there is in any other operation in surgery. I do not know anything which interferes more with the treatment of urethral affections than the apprehension, on a patient's part, of undergoing a repetition of catheterism where bleeding and pain were the chief features of the proceeding. Be careful, then, to avoid such an impression being formed, as it will seriously interfere with the carrying out of the necessary treatment. The most serious hitch that young operators experience in passing catheters, even along normal urethras, is when the point of the instrument reaches what I have called the fixed portion of the canal. On its way through the urethra, anterior to the triangular ligament, the catheter is apt, by gravity, to exercise a greater pressure on the floor than on the roof of the urethra; consequently, when it arrives at the fixed point, the instrument is below the level of the aperture in the ligament.

This is shown in Figure 8, from Dittel. If you exercise pressure, the urethra will be torn, and blood

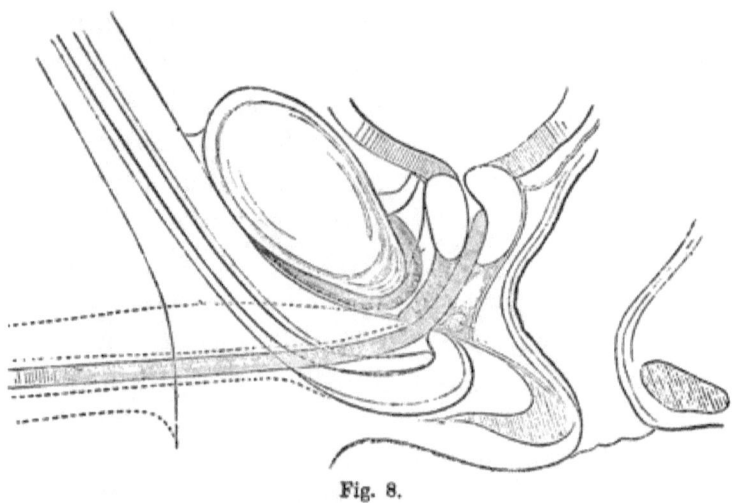

Fig. 8.

will flow. If you remember that the hitch is best avoided by keeping along the roof of the urethra, and overcome by elevating or rather drawing up the point of your instrument, difficulty need not be anticipated.

There is very considerable variety in the form and make of urethral bougies and catheters. Some you can shape for yourself according to your fancy, or in reference to the particular case, or even leave it to the urethra to mould; these include the gum-elastic instruments. Others, again, are made of silver, or some light metal, curved in accordance with the fashion prevalent in their day—fashion appearing, to some extent, to influence matters surgical as well as less important ones. Hence varieties in the curvature of metallic instruments are met with. I prefer the short curves, such as the Edinburgh instruments which are associated with the name of the late Professor Syme. I dislike operating, as one has to do sometimes in cases of emergency, with instruments curved differently from those I am accustomed to; and I would advise you, as far as practicable, to adhere to one form of instrument, if you wish to acquire dexterity in its use.

When a surgeon undertakes the treatment of a presumed case of stricture, it will be necessary for him to *explore* the urethra, and, in doing this, he should endeavour to obtain the greatest amount of information with the least amount of pain and distress to his patient.

It is my belief that the instrument which best enables us to fulfil these conditions is the plain bougie, with the end slightly rounded—not bulbous—to facilitate its introduction. If a surgeon cannot by this obtain all the information he requires about the urethra, I am sure he will not be assisted by any of the bulb or olive-headed instruments which are vaunted for this purpose. It is a matter of touch and of handling. What is wanted is the *tactus eruditus*—the acquisition of which requires quite as much practice as is needed to detect fluctuation. To the unpractised hand, the whole length of the urethra is a stricture, which no form of instrument renders otherwise.

For the dilatation of strictures a conical instrument is better, as it does its work on the principle of the wedge; for this purpose you generally see me employ

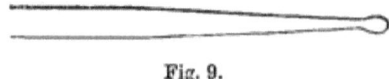

Fig. 9.

the bougie-à-boule (Figure 9). It is not so reliable as the plain instrument for exploring a strictured urethra, but it is preferable as a dilator. In selection, " preference should be given to such instruments as are rather stiff, but have a long, slender, flexible neck supporting the bulb. When held vertically, bulb

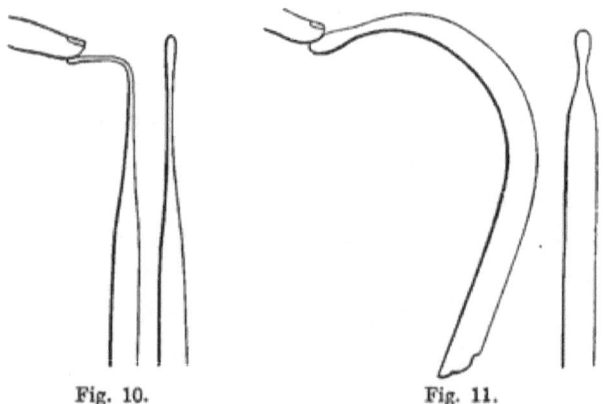

Fig. 10.   Fig. 11.

uppermost, and touched upon the olivary tip, the neck should yield at once. Such an instrument will guide itself safely, and override obstruction. The olivary points found on the English conical bougie are useless as far as any advantage derived from the bulb is concerned, from a neglect to make the neck of the instrument flexible." *

You have also seen me use with advantage, in tight, tortuous strictures, cone-shaped bougies (Figure 12) without the bulb; these, however, are rather apt

Fig. 12.

to hitch in the lacunæ of the urethra, and, consequently, are not so easily passed as the bulbous-headed instrument.

There is another form of pliable bougie which you will find exceedingly useful in the treatment of tight strictures; I allude to the filiform bougie. With care and patience it can be insinuated through the finest and most tortuous canal. The direct dilating power you can exercise upon a stricture is certainly not great; indirectly, however, the instrument is of service, and, after one or two trials, increased sizes may be made to follow. †

There is an adaptation of the filiform bougie which is not so well known as it ought to be, as in the treatment of the worst forms of stricture its use is invaluable. I refer to the tunneled bougies and

* *Genito-Urinary Diseases.* Van Buren and Keyes, p. 107.

† A very full description of these instruments will be found in a paper entitled "Filiform Bougies, their Uses and Advantages," by Dr. J. C. O. Will. *Edinburgh Medical Journal,* April, 1877.

catheters of Gouley, which excited much interest at a recent meeting of the Lancashire and Cheshire Branch of the British Medical Association, where I showed them and illustrated their use in several cases representing stricture in its worst form. The following are abstracts of the cases referred to :—

CASE I.—On June 27th, 1878, J. P., aged 27, was admitted into the Liverpool Royal Infirmary suffering from stricture of several years' duration. On this, as well as on a previous occasion, it was found necessary by my house-surgeon, Mr. Hodgson, to tap the bladder above the pubes with the aspirator. This gave immediate relief. Two days after his admission the patient again had retention whilst I happened to be in the Infirmary. I succeeded in passing one of the finest filiform bougies, and upon this a tunneled catheter, which was retained for some hours; from this date, dilatation upon the same principle was gradually commenced, the size of the tunneled instrument being increased from time to time until a No. 9 ordinary bougie passed easily when the patient left the hospital.

CASE II.—J. W., aged 60, was admitted under my care on June 12th, 1878. Thirty years previously he had a fall on his perinæum, rupturing his urethra, for which perineal section was successfully performed by Dr. Evans, of Belper. Unfortunately, the patient does not appear to have followed up the treatment of his own case by that regular introduction of bougies which in all cases of traumatic stricture is absolutely necessary, and occasional attacks of retention was the natural consequence. As I expected, I found a very tight stricture, which would only admit a filiform bougie. Upon this a tunneled bougie was passed, and dilatation continued until a Holt's instrument could be introduced. On several subsequent occasions I passed Holt's instrument, using it as a dilator on the principle of a glove-stretcher; by these means the dimensions of the urethra were soon enlarged, and the patient was able to leave the Infirmary, passing urine in a good stream.

CASE III.—W. J., aged 42, was admitted on July 5th, 1878, suffering from retention of urine, which had existed almost completely for a week, it having been found impossible to pass a catheter. On admission, I could only get into the bladder the finest filiform bougie, upon which a catheter was passed. The urine was most offensive, and cystitis was present. Rapid dilatation was employed until a catheter was passed sufficiently large to allow of the bladder being washed out. In addition to the cystitis, there was a large suppurating pouch behind the stricture, and extensive kidney-disease. The condition of this patient illustrated the consequences which may arise where a stricture of the urethra is allowed to remain untreated. He gradually sank with symptoms of uræmia, and died on July 14th. A *post-mortem* examination revealed the condition that had been predicted—viz., a suppurating pouch behind a very extensive stricture, cystitis, and suppurative nephritis.

The tunneled bougie and catheter consists of, first, a fine filiform whalebone bougie, which acts as a guide; and, second, a catheter or bougie so tunneled as to run on the guide. To use the instrument the guide is first passed through the stricture into the bladder. I hardly like speaking quite positively, but I feel very much disposed to say that this can always be done. These instruments are so fine that it is no use oiling them; you must lubricate the urethra first by injecting some olive oil. In using them remember what I said about strictures being eccentric with the opening, not in the axis of the urethra, but, so to speak, in a corner; therefore, when you meet with difficulty in passing them, slightly bend their tips and introduce them with a rotatory movement, twizzling them, in fact, between the finger and thumb when the stricture is reached. If you feel satisfied from their fixity that they have entered either a false passage or a lacuna, do not withdraw them, but pass another; if you have stopped up one gap, the next you introduce is so much the more likely to find its way into and through the stricture. You have seen me with five of these small bougies in the urethra at once, all simply serving to fill up lacunæ and old false routes; and then the sixth, having nowhere else to go, readily passes through the stricture and enters the bladder. About the latter result you can have no doubt when it is accomplished; the feel is quite unlike anything else, and a little practice soon teaches you this. These instruments, though so fine, are very tenacious, and you need have but little fear of their giving way.* On threading the bougie or catheter on to the guide, run the former slowly along the guide until the stricture is reached. Keeping to your guide, should the stricture feel hard or firm, you will be able to exercise more pressure with your instrument to overcome the stricture than you would otherwise like to do. In commenting upon my remarks at the British Medical Association, Mr. Lund drew attention to an objection that can be raised to the use of these instruments—viz., that, on passing the metallic bougie along the whalebone guide, unless

---

* Mr. Teevan records a case of a whalebone guide breaking where much force had to be applied. No ill results followed. (*Lancet*, Feb. 7, 1874.)

care is taken, the bougie is apt, on reaching the stricture, to double up on the guide, and then, if force is exercised, a false route may unintentionally be made.

Short of failing to introduce the guide, this is the only accident that is likely to occur in the use of these instruments. It does not, in my opinion, detract from their efficacy, inasmuch as I can hardly imagine any surgical instrument being made without requiring in its use that skill and knowledge which can only be acquired by experience and observation. I admit the propriety of Mr. Lund's comment, and record it as a point to be remembered.

I have described these instruments very fully, as I am sure they have only to be more generally known to be appreciated. (Fig. 13.)*

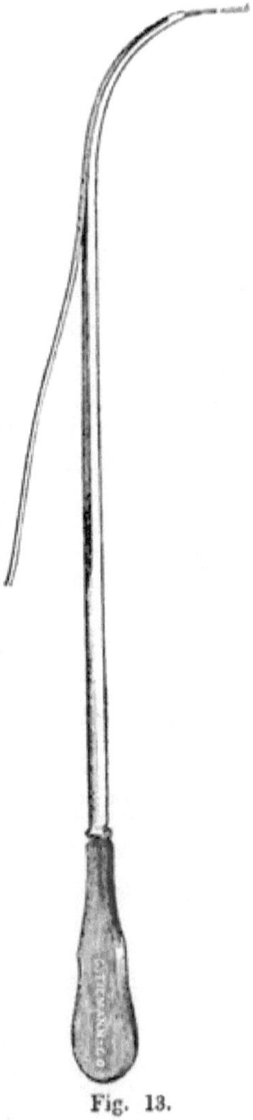

Though I believe you will generally find flexible instruments suitable, you will meet with cases where, with all your dexterity and patience, either from the hardness or tenacity of the stricture, or from the existence of a false passage, a flexible instrument will

Fig. 13.

* These instruments are made for me, after the model of Dr. Gouley's, by Mr. Wood, 81, Church-street, Liverpool.

not pass. You will then resort to bougies made of metal. Let me give you one caution in reference to them. The smaller-sized instruments, unless used with delicacy, may occasion a considerable amount of damage, as, by a very little pressure, they may be made to work their way out of the urethra. I showed a specimen, some years ago, at the Medical Society, where, in the hands of an experienced operator, a fine metallic bougie had been made to leave the urethra in front of, and to re-enter it behind, a hard stricture.

The metallic instruments generally in use are of an uniform diameter throughout, the end being slightly rounded to facilitate introduction. There is an extremely useful variety, in which the principle of the wedge is introduced, the instruments being so made as gradually to increase in size from the tip to the curve. The four I have in use are thus arranged: the first represents on the curve, 1 to 3; the second, 4 to 6; the third, 7 to 9; and the fourth, 10 to 12. It must be remembered that these four bougies represent a very considerable amount of dilating power, and are to be employed with caution.

With the instruments I have now brought under your notice, provided you will learn how to use them, I do not think you will meet with difficulty, either in detecting strictures or dilating them when necessary.

There are one or two points of which I should like to remind you. In the first place, take care that the instrument you use is smooth, warmed, and well lubricated. For the last purpose I employ castor oil in preference to anything else. Being more viscid than other oils, it is not so easily rubbed off by the first portion of the urethra. I first saw it used at St. Bartholomew's Hospital, by the late Mr. Wormald, whose dexterity with the catheter was proverbial. Vaseline is an excellent lubricant, and is much used for urethral purposes by some practitioners.

In using fine instruments, which will not carry much grease on their surface, it is a good plan to inject the urethra with oil before attempting to pass them.

Take care that all instruments passed up the urethra are most scrupulously clean. You would not like to convey bacteria, or other such-like noxious agents, into the bladder; hence a carbolized oil may be advantageously employed. There is no doubt that cystitis has been provoked by unclean instruments. I once heard of a practitioner being threatened by a person with an action for damages for giving the latter a gonorrhœa by the use of a dirty bougie. On more serious reflection, the person came to the conclusion that he might have some difficulty in sustaining his case; he therefore told me, after consulting me as to the probability of success, that being a married man he thought, under the circumstances, it would be more expedient to desist. I agreed with him.

In the employment of gradual dilatation, avoid inducing anything like extreme tension. There is nothing so repugnant to the tissues as tension, or which they are more apt to resent. I have seen a case of gradual dilatation delayed for weeks by a neglect of this precaution. In using bougies, then, in increasing sizes, stop short of the one you feel sure may be passed, but will be a very tight fit. Proceed cautiously, let a few days elapse, and you will find that, on the next occasion, the size which would have required some force now passes readily.

As a rule, four or five days should elapse between each sitting.

In the employment of flexible instruments particularly, I should advise you to measure them with a gauge, and to work systematically from this. These instruments often vary, and do not correspond with the numbers marked upon them. I prefer the French gauge, as the increase in size is much more gradual than that adopted in England.

I make use of a catheter gauge I first saw employed in America, and which is preferable to others, not only on account of its more convenient construction, but because it combines in one instrument

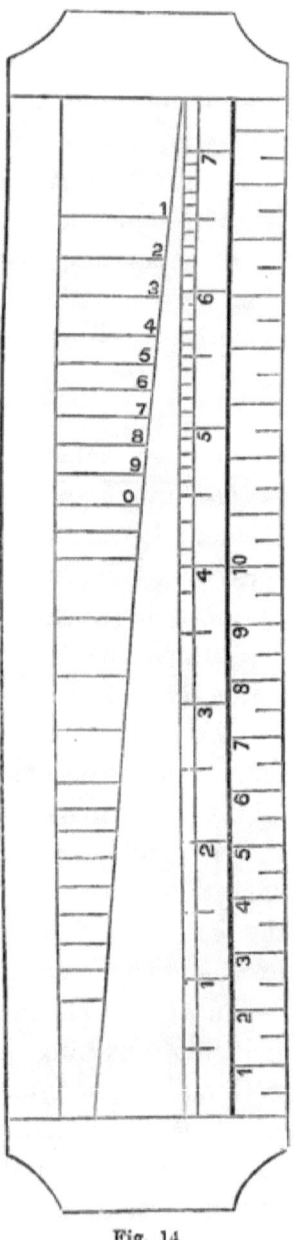

Fig. 14.

the various systems which are used for gauging urethral instruments. (Fig. 14.)

The degree to which dilatation should be practised is a point of much importance. The English make of instruments usually determines No. 12 as the limit. I prefer, as a rule, three or four sizes above this. To avoid the meatus being kept on the stretch the whole time the instrument is being passed, I have my own bougies, above No. 10, made to diminish from the curve towards the handle. When we consider that in the employment of gradual dilatation we get little more than the stretching of the urethra, it is a matter of importance to ensure that the stricture is sufficiently stretched; and this, in the majority of instances, is not accomplished by the English No. 12 bougie.

When practicable, instruct your patient how to pass a bougie for himself, as all modes of treatment require the occasional use of the instrument; and this, when he is properly educated, may generally be left to the patient, after the urethra has once been sufficiently dilated. In the case of gum-elastic bougies, it is well to caution the patient against retaining them in their service too long; they are apt to become brittle, and are then, of course, dangerous.

To patients who are very intolerant of catheterism, or in whom there is difficulty from spasm, in passing an instrument an anæsthetic may be administered with advantage. As a rule, for these cases, I prefer chloroform to ether, as, with the latter, the stage of excitement is usually more prolonged, a point worthy of consideration where we have a distended bladder. I mention this, as I prefer, for surgical operations generally, ether to chloroform. In using anæsthetics for this purpose, bear in mind Sir Henry Thompson's observation. "Let it be remembered that chloroform is administered, not for the purpose of permitting the instrument to be used with greater force than before, but in order to produce perfect anæsthesia and relaxation of the muscles."*

And now it will be proper to ask what we may expect from treatment by gradual dilatation. In the earlier forms of stricture, where the obstruction is cellular rather than fibrous, you will find that the introduction of the bougie exercises a healthy stimulus on the part, and leads to the removal of the effusion producing the obstruction.

In advanced stricture, where the adventitious deposit has been allowed to become cicatricial or indurated, I fear dilatation will do no more than dilate. You can stretch the narrowed urethra to a size corresponding with the natural dimensions of the canal, and, in the great majority of instances, by a moderate amount of care and persistence in treatment, you can keep it so stretched, but I have not been able to infer or to demonstrate that absorption, under these circumstances, actually takes place. The absorption which may have been observed to follow is to be attributed to the stimulating effect produced upon the urethra by the passage through it of what can only be regarded as a foreign body. We all know that pressure does produce absorption, but in order that

* Thompson, *On Stricture*, p. 178.

it may so act, it must be continuous, and not interrupted.

I may conclude my remarks upon treatment by gradual dilatation by observing that it is adapted to all strictures of recent date, and to those more advanced where the process of dilatation is not intolerable to the patient, and where, subsequently, the dilated condition may be sustained with a moderate amount of care.

I will now pass on to notice the treatment of stricture by what is called continuous dilatation, a method which possesses some very decided advantages.

It is based upon the observation that when a bougie is retained within a stricture sufficiently long to set up inflammatory action, the stricture yields, so that within forty-eight hours or so a bougie several sizes larger may be readily passed; by the extension of this principle, dilatation, even of the tightest stricture, may be very speedily accomplished. Hence it is well adapted to tight strictures where catheterism is attended with more than ordinary difficulty.

In gradual dilatation, the temporary pressure of the bougie stretches the stricture, and the stretching may, in favourable cases, be carried to such an extent as, in a great measure, to deprive the stricture of its contractile power; just as the frequent stretching of an elastic material, such as an india-rubber band, weakens its contractile, or rather recoiling, power.

In continuous dilatation limited inflammation of the urethra is excited, under the influence of which the stricture material melts down, so as to render the canal readily dilatable.

To carry out this treatment, a gum-elastic catheter is introduced, and retained by some suitable contrivance. A convenient way of keeping a catheter in this position is by a ring passed over the penis and secured to the body, the end of the catheter being attached to the ring. I see that Mr. Furneaux Jordan prefers a bougie to a catheter for this purpose, the advantages of the former being —"a more rapid and complete dilatation, due to the hydrostatic pressure of the urine along the exterior of the bougie. A bougie is more easily introduced than a catheter. When the finest bougie is once in, it need not be taken out—no slight boon; the ordinary acts of micturition are preserved; every kind of apparatus for keeping the bed dry, or for any other purpose, may be dispensed with." *

I have found a fine bougie—for instance, the filiform—answer all the advantages claimed for it by Mr. Jordan.

In some cases of tight stricture, it is found impossible to introduce a flexible instrument. The ordinary metallic catheter answers the purpose equally well, though it is not so comfortable for the patient, as it necessarily restricts his movements much more.

The instrument being introduced, you should retain it until it moves with complete freedom. This occupies usually from forty-eight to seventy-two hours, and is attended with varying degrees of local inflammation. When the instrument is changed, a larger one is substituted, and in this way a tight stricture may be fully dilated in the course of a few days.

* *Lancet*, January 29th, 1876.

There are two precautions which I would repeat. First, the instrument must not fit the stricture too tightly. Nature invariably resists tension. Second, the end of the instrument should only be just within the bladder. If it exceeds this, it is pretty sure to act as an irritant to the bladder; and an attack of cystitis is of course to be avoided.

It is my practice to have the temperature noted when treating a patient by continuous dilatation; a rapid rise in the thermometer indicates that the instrument should at once be withdrawn. Experience has shown me that this is a precaution which should not be neglected.

To soothe the restlessness which the restrained position sometimes occasions, I am in the habit of administering morphia in small doses at regular intervals.

In estimating the value of continuous dilatation, I have found its results compare favourably with those obtained by the gradual method. In the latter, it is only in the early stages of the disease that absorption is obviously promoted; in the later stages the stricture is merely stretched to the desired extent. In continuous dilatation I believe that not only is absorption promoted, but what is left behind of the stricture-material is less disposed to contract again.

If we could by any means, chemical or otherwise, deprive the scar-tissue, of which strictures are composed, of its tendency to contract, we should effect more for the treatment of stricture than any of the purely mechanical devices have yet done. The prosecution to a successful issue of the inquiry, By what means can scar-tissue be more nearly assimilated to healthy tissue and be deprived of its contractility? would prove of service to surgery generally, and especially to this branch of it. Working in this direction some years ago, I published a paper in which I referred to some advantages I had found in the use of belladonna as a topical application to strictures.*

As expressing views which a more extended experience has confirmed, I will quote from the article alluded to. "As a topical application, I have used it (belladonna) with benefit in some of the more obstinate forms of stricture, especially those consequent on injuries of the urethra.

"Referring back, we find some of the surgeons of the last century strongly recommending it, where spasm was present, to produce relaxation; but as we are now provided with remedies more certain in their action for combating this complication, this drug has naturally fallen into disuse.

"It is not, however, to any anti-spasmodic influence it may possess that I purpose to allude, but to the power which I believe it has of directly influencing and effecting a change in the obstructing material. I was first induced to give it a trial in these cases by observing the benefit that followed its application to cicatrices and growths of a fibrous character resembling them; in one instance, especially, an unsightly defor-

---

* "The Therapeutic value of Belladonna in some Diseases of the Bladder and Urethra." By Reginald Harrison. *Liverpool Medical and Surgical Reports*, 1868.

mity on a young woman's forehead, the constant application of the extract of belladonna produced a very decided effect; the cicatrix, though not disappearing, became much softer, more like the healthy skin around it, but what was of still more consequence, it lost almost entirely the disposition to contract, which is productive of such painful deformities. I have used it with undoubted advantage in other cases, and I therefore had reason for anticipating an equal benefit from its application to internal cicatrices that gave rise to inconvenience. Of these, the most distressing and obstinate is the stricture following injury to the urethra. It may be dilated very often, readily, but soon returns unless constantly under supervision. Dilatation simply *stretches* the cicatrix without in any way depriving it of that contractile power which is the essence of the disorder. Some patients may be instructed how to introduce an instrument for themselves, but this, for obvious reasons, is not of universal application. Others may be permanently relieved by proceedings of a more strictly operative character, and with comparatively little danger, but from them we find they sometimes shrink. It is to meet such cases that I would suggest the use of belladonna.

"The most convenient way of applying it is with the oleum theobromæ, which is sufficiently hard at ordinary temperatures to permit of its ready introduction into the urethra. I generally recommend two grains of the extract of belladonna to be used in this way twice a day, in conjunction with dilatation by bougies; the belladonna should be persevered in after the bougies have been discontinued. Very great, and I believe permanent, benefit has resulted from this plan of proceeding, and in cases where the bougie treatment alone had previously only effected a temporary relief. I may also add that my observation is confirmed by others who have given this plan a trial."

If improvement is yet to come in the treatment of stricture, it is not to be expected from mechanical means, but in the direction referred to in the quotation from the views I expressed some dozen years ago.

Those who are curious in regard to the many different ways that have been suggested for the treatment of stricture will find interest in reading how hydraulic pressure and the pressure of air, including suction, have been employed for this purpose.*

The possibility of dilating a stricture by air is referred to by Dr. Kerrison as having been utilized by the Egyptians in very primitive days for the purpose of removing stones from the bladder.† I merely refer to such suggestions and practices to say that in the case of stricture they are quite inapplicable, as their tendency, as well as that of all other forms of indiscriminate pressure, is to dilate the most yielding portion of the urethra, and that certainly is not the point where it is strictured.

The dilatation of strictures is usually effected by

---

* *Medical Record*, February, 1878. *Gazette Hebdomadaire*, No. 32, 1872. *British Medical Journal*, October 27, 1877. "Employment of Suction," by Dr. O'Connell, *Lancet*, March 2, 1872.

† "On the Dilatation of the Male Urethra by Inflation for Extraction of Calculi from the Bladder, as practised in Egypt near 250 years ago," by R. M. Kerrison, M.D.—*Med. Chir. Trans.*, vol. xii.

the introduction of instruments from before backwards; exceptions to this are exceedingly few. I have elsewhere alluded to Mr. Furneaux Jordan's suggestion,* that advantage may be taken of the dilated condition of the urethra behind a stricture to reverse this order of proceeding, and pass the instrument from the bladder side of the obstruction towards the meatus urinarius; this Mr. Jordan has effected by an incision through the rectum.

Some cases are recorded by the late Mr. Callender and by Mr. Howse † of stricture treated by the aid of supra-pubic incision into the bladder. The circumstances that would render such operation desirable are so exceedingly rare that I refer to them here rather, as Mr. Callender remarked, as being "surgical curiosities," than as commending this procedure so long as it is possible to attack a stricture from the front. As I have never met with a case where this has not been possible, and therefore have had no opportunity of demonstrating to you the operation of supra-pubic incision into the bladder, I will here only caution you against making such incisions, either for strictures or for stone, unless no other course is open. In Mr. Jonathan Hutchinson's case ‡ of supra-pubic lithotomy, the constant welling up of urine through the wound occasioned a trouble which was not by any means exceptional.

* Page 38.
† *Clinical Society's Transactions*, vol. xii. "Concerning the Operation of John Hunter in certain cases of Impassable Stricture (Supra-pubic Operation.)" Dr. MacDougall.—*Edin. Med. Journal*, 1879.
Clinical Society's Transactions, vol. xii.

I saw a man at Dr. Gouley's Clinique at Bellevue Hospital, New York, who had stabbed himself above the pubes for the purpose of relieving his retention. Great difficulty occurred in the management of the case from the constant discharge of urine through the wound. When I last saw him his condition appeared most unpromising.

SIXTH LECTURE.

URETHRAL FEVER — SUPPRESSION OF URINE — HÆMORRHAGE FROM THE URETHRA — FALSE PASSAGES.

No operation in surgery, however slight or dexterously performed, can be regarded as absolutely free from danger. Nature will, at times, resist interference, however loudly demanded, in a manner little anticipated, and even inexplicable. In this "chapter of accidents," I propose first to draw your attention to certain phenomena occasionally following the passage of instruments along the urethra, to which the term "Urethral Fever" has been applied.

It not unfrequently happens, after a bougie has been passed, that the patient experiences a sense of chilliness: this "rigor," as it is technically called, is usually followed by more or less febrile excitement, which subsides without occasioning the patient much distress.

Passing to the opposite end of the scale, we now and then meet with a case where the most alarming symptoms of shock are speedily followed by death.

I very well remember, as I happened to have charge of the ward at the time, the fatal case alluded to in Mr. Banks' excellent pamphlet on Urethral Fever,[*] where death followed the introduction of a

[*] *Edinburgh Medical Journal*, 1871.

bougie in six and a half hours. Fortunately, such tragical instances as these are exceedingly rare.

Between the two extremes I have instanced we meet with every degree of severity, but in all a resemblance to fever can be made out. I do not think I can better inform you about Urethral Fever than by narrating to you, in detail, the particulars of a case which, having been recently under my care in the Infirmary, many of you had an opportunity of seeing. Those of you who saw the case with me will remember the extreme severity of the symptoms, and the small amount of hope we entertained of saving the patient's life. Fortunately, however, a fatal termination was avoided; and this happy result was, in a large measure, due to the assiduous and unceasing care of those who watched the patient throughout his illness.

T. R., aged 40, an engineer, was admitted into the Infirmary on September 30th, 1876. For some time previously he had suffered from stricture, the distress from which had recently much increased. I found a stricture in the membranous portion of the urethra, which would only admit a No. 3 bougie (English). Progress in treatment was very slow, the stricture being unusually hard to dilate. The urine, though passed in a small stream, was healthy. On Oct. 20th I introduced a Holt's dilator, but as the instrument hitched at the stricture and would not pass into the bladder, I postponed a further attempt. Some slight bleeding took place, which speedily stopped.

Four hours after this the patient was seized with a rigor. I should say that on no previous occasion had this occurred. At 9 p.m., or three hours after the rigor, his temperature had risen to 104° Fahr.; at 12 (midnight) he vomited. At 5 a.m. on the following day (21st) he had another rigor, succeeded in an hour by vomiting, the temperature being 104° Fahr. On the occurrence of the first rigor, five grains of quinine were given to him, and continued every six hours for three doses.

On the next day (22nd) he had three rigors at intervals, followed by vomiting, the temperature varying between 95° and 102°. At 5 p.m. on this day he, for the first time since the introduction of the instrument, passed urine. It amounted to only four ounces, was high-coloured, and contained much albumen, depositing half on boiling. He had, therefore, remained for over fifty hours without passing urine, and four ounces represented the extent of the excretion during this period.

On the 23rd his temperature varied between 94° and 96·2°. He was exceedingly drowsy. The tongue was dry and brown, and the pulse quick and thready. There was in fact every indication of speedy death. At 10 a.m. on this day he passed two ounces of urine. The next day his temperature varied between 94° and 100°. He passed nine ounces of urine, and vomited once. His general condition remained much as on the previous day. The skin was dry and of a dirty-sallow colour.

On the 25th he passed twenty ounces of urine during the twenty-four hours; on the 26th, eighteen ounces. The albumen had now fallen to $\frac{1}{12}$th. On the 27th, the temperature varied between 95° and 100°. The urine was normal in quantity, but still contained albumen. At this date convalescence commenced, and by the 30th of October all grounds for uneasiness had disappeared, the urine being now normal, both in quantity and quality. Nutrient enemata were given whenever the patient was unable to take food in consequence of sickness; stimulants were also administered according to circumstances.

For the first four days the patient remained in a semi-comatose condition, occasionally waking up and talking, or rather muttering, deliriously. This state of mind continued until the normal secretion of urine was re-established. The bowels were constipated throughout, and from time to time were relieved by enemata. He appeared to vomit on the occasions mentioned, quite independently of any food he might have been taking. I was sorry that the vomited matter was not kept for closer examination, as this discharge might have been found to contain material which should have been eliminated by the kidneys.

(T. R.) CASE OF URETHRAL FEVER.

Date.	Time.	Temperature.	Remarks.	Urine Passed.
Oct. 20. (Friday).	6 p.m.	103	Rigor.	
	9 ,,	104	Quinine 5 grs.	
	10 ,,	103	...	None.
	12 ,,	103·2	Vomiting.	
Oct. 21. (Saturday).	2 a.m.	103		
	3 ,,	102·3	Quinine 5 grs.	
	5 ,,	104	Rigor.	
	6 ,,	104	Vomiting.	
	7 ,,	103		
	8 ,,	102·4		
	9 ,,	100·3	Quinine 5 grs.	
	10 ,,	102	...	None.
	11 ,,	101·3		
	12 ,,	101		
	2 p.m.	100		
	4 ,,	99		
	6 ,,	98		
	8 ,,	96		
	9 ,,	97		
	12 ,,	95		
Oct. 22. (Sunday).	6 a.m.	96		
	1 ,,	100	Rigor.	
	2 ,,	100·2	Vomiting.	
	3 ,,	96		
	8 ,,	96		
	10 ,,	96	Vomiting.	
	12 ,,	97	Rigor.	

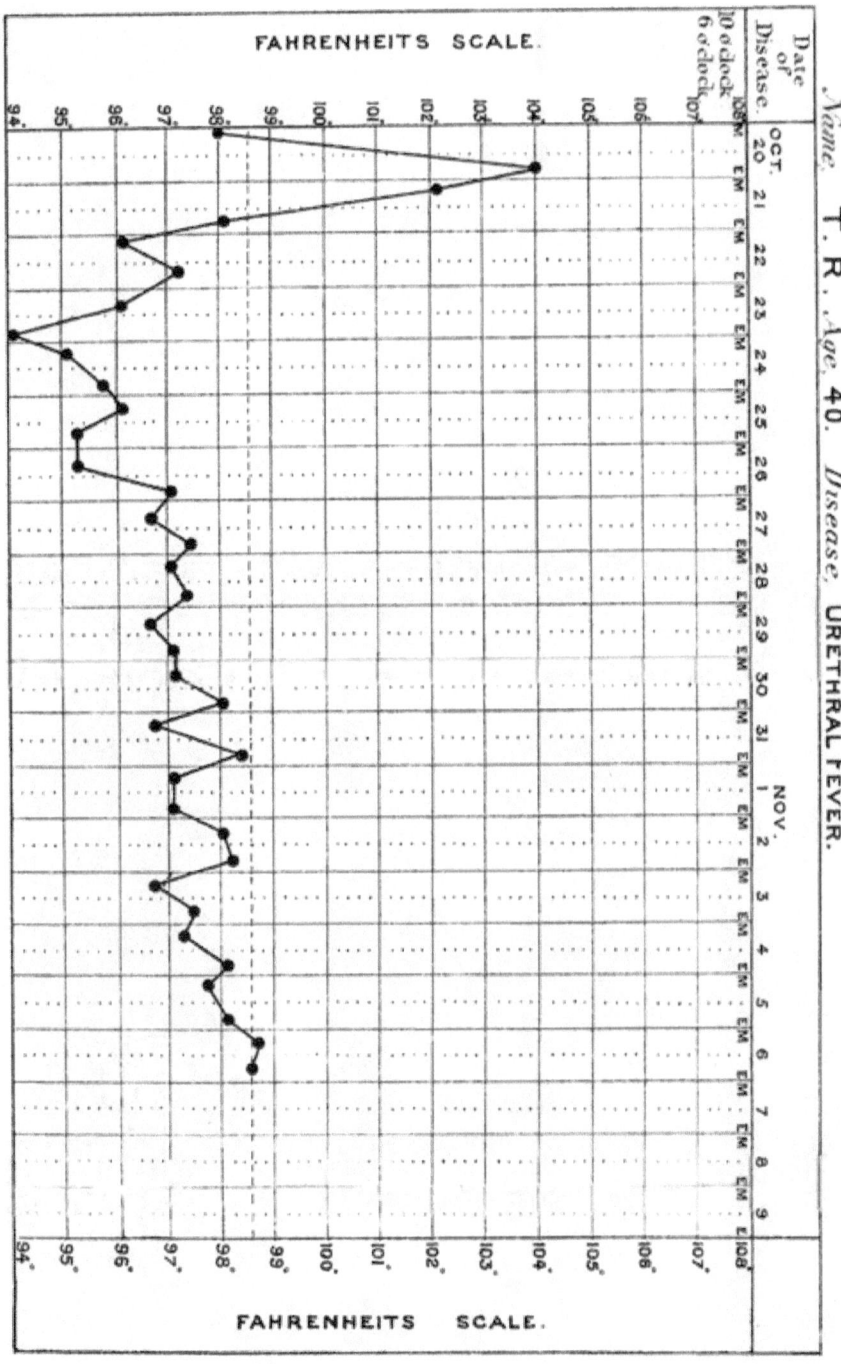

93

Date.	Time.	Temperature.	Remarks.	Urine Passed.
Oct 22. (Sunday).	2 p.m.	98	Vomiting.	
	3.30 „	101	Rigor.	
	4.30 „	102	Vomiting.	4 Ounces;
	5.30 „	97·2	...	½ Albumen.
	7 „	100		
	9 „	100		
	11 „	97·4		
	12 „	96·4		
Oct. 23. (Monday).	2 a.m.	95.2		
	4 „	95		
	6 „	95		
	8 „	94		
	9 „	94·2		
	10 „	94	...	2 Ounces.
	11.30 „	94·4		
	1 p.m.	94·4		
	2 „	94·4		
	4 „	95		
	5 „	95		
	7.30 „	94		
	9 „	95		
	12 „	96·2		
Oct. 24. (Tuesday).	1 a.m.	98		
	3 „	100	Vomiting.	2 Ounces.
	5 „	95		
	7 „	95		
	9 „	95		
	11 „	94·1	...	1 Ounce.
	1 p.m.	95	...	2 Ounces.
	3 „	95·2		
	5 „	95	...	4 Ounces.
	9 „	97·1		
Oct. 25. (Wednesday).	2 a.m.	96		
	4 „	94		
	6 „	94		
	9 „	96		17 Ounces.
	12.30 p.m.	94		3 Ounces.
	4 „	95·4		
	9 „	95		

Date.	Time.	Temperature.	Remarks.	Urine Passed.
Oct. 26. (Thursday).	2 a.m.	97	...	18 Ounces. $\frac{1}{12}$ Albumen.
	4 "	96		
	6 "	94		
	9 "	95		
	12 "	96·4		
	5 p.m.	97		
	9 "	97		
	12 "	101·3		
Oct. 27. (Friday.)	2 a.m.	101·3	...	Normal in quantity.
	3 "	98		
	9 "	96		
	12.30 p.m.	96	...	Albuminous.
	5 "	97·4		
	9 "	96		
Oct. 28. (Saturday).	1 a.m.	95·4	...	Albuminous.
	3 "	96·4		
	9 "	97		
	12.30 p.m.	97·2		
	5 p.m.	97·2		
	9 "	97·2		
Oct. 29. (Sunday.)	3 a.m.	96	...	Albuminous.
	9 "	96·4		
	12 "	96·3		
	5 p.m.	97		
	9 "	97		
Oct. 30. (Monday).	7 a.m.	96	...	Normal; no Albumen.
	9 "	97		
	12 "	97		
	5 p.m.	98		
	9 "	98·2		

Such, then, is an outline of this very interesting case. That these symptoms were occasioned by the surgical interference to which the urethra was subjected no one can doubt, nor can we come to any other conclusion than that they were the immediate effect of shock propagated through the sympathetic system of nerves, which so largely supplies the generative organs.

In this case there was no evidence of previous disease of the kidneys; the effect, however, produced upon these organs during the existence of the symptoms was very remarkable, excretion for a time being almost entirely arrested. As bearing upon the view taken that urethral fever is produced by nerve shock, I would mention that, on referring to my notes of those cases where anæsthetics have been used for catheterism, I find an almost complete absence in them of such symptoms, thus indicating that pain is an important factor in their causation. There is no kind of pain so likely to produce shock as that resulting from tension, as when a tightly-fitting instrument is slowly forced through a strongly-resisting contraction.

Persons who are suffering from kidney disorder, which is a frequent concomitant of stricture, are undoubtedly more liable to urethral fever than those who are not. Such persons feel the effects of shock much more acutely than others, and consequently, in these cases, we must limit any surgical interference to that which is absolutely necessary for the preservation of life, regardless of other considerations. We are again reminded of the importance of making a careful examination of the urine in all cases in which we are about to undertake the treatment of stricture. No cases are so much benefited by the observance of the medical and hygienic preliminaries usually employed with patients about to undergo a surgical operation as these, and when circumstances are not pressing, they should not be dispensed with.*

In the treatment of urethral fever, I cannot but endorse the favourable opinion which has been expressed as to the efficacy of aconite and quinine. The former is almost a certain prophylactic when administered in two-minim doses of Fleming's tincture immediately after catheterism, as first suggested by Mr. Long.† On the latter, reliance must be placed on the development of the symptoms.

The most serious aspect of urethral fever has reference to the state of the kidneys and the excretion of urine. In the slighter cases, there are no appreciable changes in the urine; but in the severer forms, it may be albuminous, sanguineous, or even completely suppressed. Whatever views may be held as to the precise pathology of urethral fever, there can be no doubt that it is accompanied with more or less renal engorgement, and sufficient, by mechanical pressure, to account for the altered changes in the urine just referred to. I cannot help thinking, from a study of urethral fever as seen following operations on the urethra, that we must look for an explanation of its phenomena in renal thrombosis, more or less complete, as the initial lesion. Whether this be true or not, I feel sure that the severity of the complication may be

---

* "I prepare all cases—with the exception, of course, of urgent cases—for five or six days before they are placed under mechanical treatment, and consequently now have in my own practice but few cases of urethral fever to treat." —Gouley, *Diseases of the Urinary Organs*, p. 38.

† *Liverpool Med. Chir. Journal*, January, 1850.

measured by the extent to which the kidneys suffer as determined by their excretion, and that the remedying of it lies in the restitution of this function. This symptom of suppression is best met by acting upon the skin, and increasing the perspiration by the employment of vapour baths, or, where this cannot be done, by placing the patient for a few minutes in a hot bath, and then enveloping him in blankets. Dry cupping over the kidneys may also be advantageously employed. For its diuretic action, Gouley recommends that a tea-spoonful of the infusion of digitalis be given every hour or two, the effect on the circulation being closely watched. The infusion is preferred to the tincture or the extract, as being a more effective diuretic. Since seeing this observation by Gouley, I have tried digitalis in two cases of urinary suppression after catheterism, one in private practice, and the other in the Infirmary. In each instance the effect was most marked, and drew forth comment.

Prevention is better than cure; and in no instance is this truth more aptly illustrated than in the disorder now under notice. If, after every operation at all likely to cause urethral fever, a state of diaphoresis were set up and maintained for a time by hot blankets and a suitable temperature, we should hear but little of urethral fever, and that little in a very mitigated form. This is my experience. Whether this consideration has any bearing upon the pathology which I have advanced, viz., that the symptoms constituting urethral fever are indicative of a state of more or less complete renal thrombosis, is a point, I think, worthy of your attention.

Hæmorrhage from the urethra is not an infrequent consequence of catheterism. When slight, it need not occasion any anxiety. In cases of granular urethritis it is difficult sometimes to pass an instrument without rupturing some of the minute bloodvessels, which, in this state, remain in a highly congested condition. I saw, not long ago, in consultation, a case of hæmorrhage from the urethra, after catheterism, of an unusually persistent character. So exhausted was the patient that it became a question whether it would not be necessary to lay open the perinæum, and expose the urethra at the place where it was supposed to be injured. Before seriously entertaining this proposition, I suggested that the subcutaneous injection of ergotine should be tried. After two injections the hæmorrhage entirely ceased.* This patient appeared to have come from a family of "bleeders," a brother and a sister having died from persistent hæmorrhage following trivial injuries. I have elsewhere illustrated the value of matico as a hæmostatic. When the hæmorrhage is free and continuous, there is reason to believe that some laceration has been occasioned to the walls of the urethra, and that a false passage has been made. The damage that is inflicted in this way is sometimes very extensive. Not long ago, there was, in my ward, a case in which I had unusual difficulty in passing a catheter for urgent retention of urine, in consequence of a false passage opening into the rectum, which had been made, a few hours previously, by a notorious secret-

---

* In the *American Journal of Medical Science*, of July, 1877, Dr. Boyland draws attention to the great efficacy of Ergot in urethral hæmorrhage.

disease quack. After I had once succeeded in relieving the bladder, no further catheterism was necessary, and I delayed the passing of bougies for a fortnight. Fortunately, by this time, the false passage had closed, and treatment was proceeded with as usual.

When a false passage is made, it is usually along the floor of the urethra, the operator allowing the point of his instrument to drop, and then, when a hitch occurs, as the more fixed portion of the urethra is reached, exercising force. Now, as the deeper portion of the urethra can be readily explored by the finger in the rectum, whenever difficulty arises here, or there is reason to believe that the instrument has deviated from its proper course, this means of assistance should not be forgotten. For, with the finger in the rectum, the hitch may be overcome, or the wrong position of the instrument detected.

When a false passage has been made, it is better to suspend further instrumental treatment for some days; there is but little risk of extravasation of urine occurring, as the direction of the laceration is contrary to that of the stream of urine. On resuming treatment, care will have to be taken to avoid the site of the false passage, as, by repeatedly opening it up, it may be converted into a sinus. As I have already stated, false passages are usually made in the floor of the urethra. I pointed out an exception to this in the case of a patient who had an old false passage leading from the upper wall of his urethra; this, however, was so far satisfactorily explained to me as having been caused by the patient in using an instrument like a stylet, which he occasionally employed at sea whenever he considered that his stricture required " breaking down." I could pass a small elastic bougie into the false route, whilst another, at the same time, could be made to traverse the stricture into the bladder. This case was satisfactorily treated by Holt's operation.

SEVENTH LECTURE.

Retention of Urine—Catheterism—Impassable Stricture—Aspiration of the Bladder—Tapping—Cock's Operation—Forcible Catheterism.

Of all the operations in surgery, there is none, perhaps, that affords such immediate relief, or calls forth greater gratitude from the patient, than the successful use of the catheter for retention of urine.

This is the most distressing accident that can happen to a patient who is the unfortunate subject of stricture. The circumstances which bring this about hardly require further notice here. Usually it is spasm superadded to organic stricture that occludes the urethra, and converts difficult micturition into complete retention. An excess of some sort is generally the exciting cause. In elderly persons especially, an attack of retention is often brought about by what is commonly called "catching cold," when the action of the skin is arrested, and a greater amount of work is thrown on the kidneys. I need hardly remind you that a person may be passing urine while he is, to all intents and purposes, suffering from retention; that is to say, his bladder is distended with urine. I have known this condition escape recognition, and the dribbling of urine ascribed to incontinence or paralysis of the bladder. The diagnosis of retention is usually so simple that it is sufficient merely to mention this to guard you against falling into such an error. Circumscribed distension above the pubes, and pressure on the rectum, serve to indicate to the touch that of which his sensations make the patient only too conscious. In judging of the degree of distension, we must to some extent be guided by the sensations of the patient. If one man is so insensitive as to require relief by the catheter only, when the fundus of his bladder reaches his umbilicus, we must not infer that another less needs it because this line has not yet been, in his case, attained. Our powers of endurance in this respect are very different. The patient with the small, contracted bladder, from long-standing stricture, suffers all the horrors of retention long before the limit I have indicated has been reached.

The consideration of treatment may be simplified very much by dividing the subject into two heads: first, where a catheter can be passed; and, second, where it cannot, or its use is impracticable. Where retention is urgent, it is undoubtedly the duty of the surgeon at once to attempt catheterism. If he succeed, relief is immediate; and attention will subsequently be turned to the removal of the cause. Where retention is not urgent, a hot bath and a full dose of laudanum frequently produce the desired effect. A very opposite plan—viz., the introduction of a piece of ice into the rectum—is considered by some an almost infallible remedy. I believe this was first suggested by Cazenave. Where there is much difficulty in getting through the stricture, Dr. P. H. Watson's steel probe-pointed catheter will be found exceedingly useful.* Its chief advantages are that, being made of steel, it is thoroughly rigid, and therefore under the absolute control of the operator; and, further, it possesses all the excellence of the smallest catheter with, from its probe-pointed extremity, all the facility of introduction presented by the probe-pointed bougie. The most recently-made instrument gradually increases in size from the bulbous extremity towards the handle, as shown in Figure 15. Before concluding that it is impossible to pass a catheter, I would advise you to try one of the whalebone filiform bougies and a tunneled catheter as described in a previous lecture. I have often succeeded at the first trial after attempts with ordinary catheters have failed and I have been requested to tap the bladder.

Turning to those cases where catheterism is impracticable—instances of which in the present day are fortunately exceedingly rare—we find ourselves provided with various expedients. Some have only to be mentioned to be condemned, whilst others have stood the test of experience. Aspiration has been very advantageously employed in this class of cases. I

Fig. 15.

* *Edinburgh Medical Journal*, July, 1869.

have selected the following as exemplifying this mode of treatment :—

CASE I.—In the summer of 1876, W. F., aged 44, was admitted into the Royal Infirmary, under my care, suffering from a stricture, the result of an injury to the perinæum. On several occasions he had been unable to pass water, and much difficulty was always found in relieving him by catheterism.

During the last few months his stricture had become much tighter, so much so that the urine only escaped in drops.

I saw him at the time of his admission; his bladder was largely distended, and he was in urgent want of relief. On looking at his perinæum, I found an old scar, and around it considerable induration. I endeavoured to pass a small catheter, but without avail. As the abdomen was exceedingly tense and the patient in great distress, without further delay I punctured with the aspirator immediately above the pubes, and removed, I should think, three pints of urine. In the course of two hours I had the patient placed in a warm bath, and afterwards gave him a full dose of laudanum. The urine began again to issue in drops, and no further retention was experienced. I kept the patient in bed for four days, during which period no attempt was made to pass a catheter, nor was the procedure necessary. It was interesting to notice how, under the influence of rest, alkalis, and purgatives, the patient's powers of micturition improved. I subsequently gradually dilated the urethra until I could pass Holt's instrument, by means of which I ruptured the stricture. In the course of three weeks the patient left the Infirmary, being able to pass for himself a No. 12 bougie.

CASE II.—W. W., a sailor, was admitted on April 24th, 1874, into the Liverpool Royal Infirmary, under my care, for retention of urine. The patient had suffered from stricture for three years, and for some time prior to his admission the stream of urine had been diminishing in size. Twelve months ago, when at sea, he had suffered from retention, and was with difficulty relieved by catheterism. Two days before his admission he had been drinking freely, and on his arrival at the Infirmary he had not passed urine for some twenty-four hours.

On admission the bladder was largely distended. Catheterism was ineffectually tried. When I saw him, shortly after admission, I found him in great distress. I attempted to introduce a catheter; but, from the state of the parts, I felt convinced that the instrument could not be made to enter the bladder without exposing the patient to injury by a persevering and perhaps protracted attempt to relieve him in this way. I then introduced into the bladder above the pubes one of the smallest needles of the aspirator, and removed a large basinful of highly-coloured urine. The patient was at once relieved. I gave him a dose of laudanum, and during the night he commenced to pass urine naturally in a small stream. A brisk cathartic was prescribed on the following day.

On the fifth day after his admission, without much difficulty, I passed a No. 3 bougie through a tolerably long and tight stricture. From this date gradual dilatation was employed, and on the 12th of May, when he was made an out-patient, dilatation had proceeded as high as No. 8. I should add, that he suffered no inconvenience from the supra-pubic puncture made by the needle, nor could any mark be discovered forty-eight hours afterwards.

It may be asked, was it impossible to introduce a catheter? I would not like to admit this in any case; for, assuming an average amount of dexterity, such an operation, in the greater number of strictures, is a matter of perseverance only. But what may be possible may not at the same time be expedient. In the second case a reasonable trial had been first made by the house-surgeon, but without avail; a warm bath and an opiate, pending my arrival, were also ineffectually tried. I was not surprised, on introducing a catheter as far as the obstruction, at this want of success, the stricture being dense and unusually hard and resisting. As the patient required immediate relief, the aspirator was resorted to in preference to the older plan of puncturing the bladder by a trocar and cannula above the pubes or through the rectum.

It may be objected that the aspirator would only afford temporary relief, inasmuch as the urethra was obstructed. To this I reply, that in the great majority of cases it is the spasm which, superadded to the stricture, determines the retention, and if temporary relief is afforded, the power of micturition becomes re-established. Here the patient, in the course of a drunken debauch, had distended his bladder to a degree over which he could not exercise a proper expulsive effort. The bladder being artificially emptied, the patient's distress was at once relieved, and on the collection of water again in the bladder he took care that it should not remain there to exceed the limit beyond which he was incapable of exercising successfully expulsive power. Time was thus allowed for getting the patient into a condition suitable for further treatment, and on the fifth day, as was predicted, the first step in the treatment by gradual dilatation was commenced, and uninterruptedly continued to a satisfactory issue. I would remark, in passing, that the treatment by gradual dilatation was carried on more rapidly in this case than I could have wished, in consequence of the patient being very desirous to resume his work. Experience shows that dilatation, to be successful, should be very gradually employed.

As a means for relieving retention of urine arising from organic stricture, pneumatic aspiration cannot fail to be exceedingly valuable; for, apart from the considerations I have urged, it is obvious that a stricture is never improved by anything like a prolonged effort at catheterism. Any laceration of the urethra (and where there is hæmorrhage this must to some extent occur) necessitates a corresponding cicatrisation, and this, by its subsequent contraction, adds to the obstruction. In tight strictures, with retention and a distended bladder, the difficulty in introducing a catheter is undoubtedly greater than where the bladder is capable of acting, and with this difficulty the risk of doing harm with the catheter is proportionately increased. The aspirator will in such cases be found a suitable means for tiding over that period of time when the difficulty is greatest, thus enabling the practitioner to commence his treatment under more favourable circumstances.

A considerable number of recorded cases show the safety with which the aspirator may be used. With a fully distended bladder it is almost impossible to injure the peritoneum; and if the finger is for a moment firmly pressed above the pubes before the instrument is introduced, until a pit or depression is formed, the passage of the needle is absolutely painless. My observation would quite confirm the remark of a patient recorded by Dr. J. Bell,* whose case has been published, "that it was the easiest way of having the water drawn off he had ever experienced."

That aspiration may be repeated an almost indefinite number of times is evident from a case recorded by Mr. W. Brown, where, for retention from an enlarged prostate, it is stated, "we used the aspirator daily, and on some occasions the pain was such as to require the operation to be performed twice in the day. Altogether we performed the operation fifteen times, with immediate relief on every occasion, and without the smallest inconvenience or injury from the punctures or perforations of the needles." †

Sufficient evidence, I think, has now been adduced to show that in the aspirator we have a valuable addition to our resources for the treatment of the class of cases I have illustrated.

In tapping the bladder with the aspirator there are two points which should be remembered: (1), only to use a very fine needle, and (2), in introducing the needle above the pubes, to keep as close as possible to the bone, and so avoid puncturing the peritoneum. The point at which the peritoneal reflection takes place I have found to be slightly variable, but a clear space of at least half an inch always exists where the needle may be introduced with safety. This space is increased when the bladder is largely distended, as in cases of extreme retention.

It has been objected by Dr. W. Macfie Campbell,

---

* *Edinburgh Medical Journal*, April 1874.
† *British Medical Journal*, May 23, 1874.

that aspiration may be followed by urinary extravasation along and around the puncture of the aspirator trocar; and he records a case where such a sequence took place.* Tapping by the rectum, which he seems to prefer, is by no means free from such an accident, and I cannot help thinking, from some experience of both methods, that if a comparison were made, say between a hundred cases of each proceeding, the balance of advantage and of risk would largely favour aspiration. I am glad of the opportunity of recording this objection, as well as the caution accompanying it, that, to prevent such an accident occurring, should the bladder again require emptying in this way, aspiration must be resorted to before anything like extreme tension of the walls of the viscus has been reached.

I would just add a remark as to the best kind of aspirator, as I have now employed the instrument on numerous occasions and for a variety of purposes. I much prefer the instrument in which the reservoir and the exhausting syringe form two distinct pieces, as the simpler looking syringe-aspirator is very apt to get out of order, and, by leaking and ejecting the fluid between the rod and its socket as the piston is worked upwards, to inconvenience the operator. The former instrument I have found free from this objection, and in all respects quite worth its rather greater cost.

Tapping the bladder with a trocar and cannula above the pubes or through the rectum are expedients which, like aspiration, are only to be resorted to where there

* *British Medical Journal*, Feb. 21, 1880.

is a prospect of re-establishing within a short period the natural passage of the urethra.

The chief objection to the older methods of tapping by the rectum or above the pubes is that they require the retention of the cannula within the bladder for some days; this is not only a source of discomfort and irritation to the patient, but is not free from the chance of infiltration of urine occurring and of permanent fistulous communication between the bladder and the rectum. Hence, as a temporary expedient for affording relief to the distended bladder, I think aspiration is to be preferred.

Where the chance of restoring the urethra is only remote, as in some old-standing cases of stricture, Cock's operation " of tapping the urethra at the apex of the prostate, unassisted by a guide-staff;" may be resorted to with advantage. I frequently see a patient upon whom I performed this operation some ten years ago. For five years previously he had endured all the vicissitudes that could happen to the subject of, at times, an impassable stricture. Since the operation he has enjoyed perfect health and comfort at the expense of micturating through the perinæum. Cock's operation is so well known that I need not further refer to it.*

It is the practice of some surgeons, whilst relieving by incision the pressing symptom—viz., the retention —at the same time to effect a division of the stricture. Such a proceeding, always difficult of accomplishment, is not to be recommended, it being better to reserve

* *Guy's Hospital Reports*, 1866.

the treatment of the stricture until the retention has been relieved, unless, as sometimes happens, after the urethra behind the stricture has been opened with the knife, a guide can be passed into the bladder.

"Forcing the stricture" is a proceeding so likely to be attended with disastrous consequences as not to be entitled to recommendation. If you cannot pass a catheter by the exercise of a legitimate amount of firmness and tact, you are pretty sure to do harm by such hap-hazard manipulations.

In conclusion, let me urge the importance, in all cases of retention, of making a careful and well-directed effort to give relief to the patient in the most effectual and speedy manner—namely, by the introduction of the catheter. In but a few will you fail; it is only after such a trial as this has been made that you are justified in entertaining the other proposals to which I have alluded.

Retention of urine from enlargement of the prostate will be considered on a future occasion.

EIGHTH LECTURE.

INTERNAL URETHROTOMY — SELECTION OF CASES — OTIS'S VIEWS — VARIOUS METHODS OF PERFORMING INTERNAL URETHROTOMY — THE USE OF OVAL BOUGIES — HOLT'S OPERATION.

THE division of strictures from within the urethra—internal urethrotomy—is a mode of treatment which has long been practised, and as at present performed in properly selected cases may be regarded as both certain and safe. In no department of surgery has greater ingenuity been shown in the devising of instruments than in this, and to attempt anything like a full description of the various urethrotomes in use would here be foreign to my purpose. I shall therefore content myself with drawing attention to those proceedings which experience has shown me to be best adapted to the purpose in view.

Much discrimination is required in the selection of cases suitable for internal urethrotomy, no operation in surgery having been brought more into disrepute by a neglect of those conditions upon which its success essentially depends than this.

The cases which are suited to this plan of treatment include those contractile forms of stricture which are not only tedious and painful to stretch by gradual dilatation, but which speedily contract on the suspension of treatment. Such strictures as these are almost invariably annular—that is to say, they embrace the entire circumference of the urethra. They may, as you have seen, be temporarily remedied by gradual dilatation, but they speedily contract and return with all their difficulties unremoved. These, as a rule, are permanently relieved and rendered manageable by internal urethrotomy.

There is a form of stricture—musculo-organic I call it—in which, after each dilatation by bougies, even when the larger sizes are reached, the operation is almost invariably followed by retention, a circumstance which is very disappointing to both the surgeon and the patient. In these I have allowed the stricture again to contract, and then treated them by internal urethrotomy and dilatation, thus combining section with stretching, a principle which is best exemplified in the treatment of contracted limbs by tenotomy, followed by extension with a mechanical contrivance. In this way these cases are invariably benefited by internal urethrotomy, and as the immediate relief is great, the risk is found to be small.

Again, experience teaches us that contraction of the meatus, and penile strictures generally, are best remedied by division with the knife; and this is in a large measure due to the precision, arising from their position, with which their section can be accomplished.

Further than this, Dr. Otis, of New York, has considerably extended the field, for he practises this plan of treatment in the earliest forms of urethral obstruction, hoping thereby to effect a permanent cure of the disorder.*

Dr. Otis very properly points out that nearly all contractions of the urethra are due to inflammation, and that "gleet is the signal which nature hangs out to call attention to the fact that the urethra in which gleet occurs is always strictured at some point." Subject to the reservation that a state of congestion, such as we have in gleet, as shown by endoscopic examination, necessarily lessens to some extent the calibre of the canal, we may admit the truth of this. Moreover, we may agree with Dr. Otis, that there is some variety in the dimensions to which the urethra may be stretched, either as in the act of micturition or in the receiving of a bougie, just as there is a variation in the size and shape of the penis.

Based upon these conclusions, Dr. Otis proposes a method of treatment, which consists in the division of such portions of the urethra as do not come up to the standard of the "individuality," this being determined by a very ingeniously devised instrument called a urethrometer; subsequently the proper calibre of the canal is maintained by the use of dilators.

Now, whatever the ultimate good results may be, I hardly think any one of us would like to repeat the experience of Mr. Berkeley Hill.† "I may state that all the cases operated upon here were those of long-stand-

---

* "The Treatment of Stricture of the Urethra," by F. N. Otis, M.D., *British Medical Journal*, Feb. 26th, 1876.

† *The Lancet*, April 8th, 1876.

ing gleets, with contraction in one or more parts of the spongy urethra, and had undergone multifarious treatment. The number of patients is sixteen: fifteen of my own and one of Dr. Otis. In five cases the gleet stopped after the operation, and the patient was, at the last report—taken, in none less than three weeks, in most some months, after the operation—able to pass a bougie of the estimated size of the urethra. In short, they may be claimed as cures. But of these five the operation was serious to two; one had free bleeding for three days; the other three attacks of rigors; of the remaining eleven, among whom Dr. Otis's own operation must be included, the gleet persisted in all; in several the urethra shrank again to its size before the operation, and in some very serious complications ensued; in four bleeding lasted several days, and in one was even alarming."

My own observation is certainly not favourable to Dr. Otis's method. I have already expressed my belief that stricture in its earliest form is curable, provided that dilatation by bougies is sufficiently carried out. Hence I reserve for internal urethrotomy cases which are not likely to be benefited by dilatation, and these, so far as my experience goes, do not include strictures in their early stage.

In contractions of the meatus of the urethra, division may conveniently be accomplished with a blunt-ended tenotomy knife. I usually make two incisions, one on either side of the median line, directed somewhat obliquely. In strictures further down the canal, but still within the penis, division may be performed with the *bistourie caché*, or with an instrument which

I have found very useful for this purpose, viz., a probe-ended knife; this may be readily manipulated within the stricture, and its division completed. (Fig. 16.)

In the deeper portion of the urethra, division is not so readily accomplished; for its performance we are provided with two sets of instruments — viz., those cutting from before backwards, of which Maisonneuve's and Teevan's urethrotomes are examples; and those cutting in an opposite direction, amongst which I may mention Watson's and Civiale's. Of these instruments I certainly find that Watson's is the best.* In principle it resembles the staff used by Syme, to which is added a knife concealed in the narrow portion of the guide, immediately in front of the shoulder, which indicates the position of the stricture to be divided. The protrusion of the knife is regulated by a screw at the handle, and by means of a dial attached to the screw the degree to which the knife is projected is at once indicated. (Fig. 17.)

One of the last cases I operated upon is a good example of this plan of treatment:—

J. H., a seaman, aged 40, was ad-

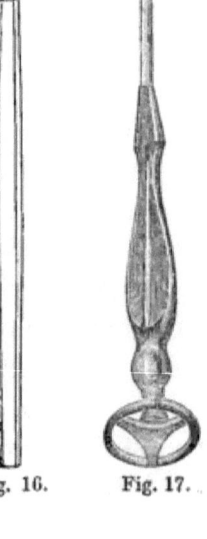

Fig. 16.   Fig. 17.

* *The Lancet*, Oct. 23rd, 1875.

mitted into No. 1 ward in September, 1877, for stricture, from which he had suffered for some years, and which had been treated on several occasions by gradual and continuous dilatation. The stricture was situated at the bulb, and would only admit a No. 1 bougie. Dilatation was an exceedingly painful process, the passage of the bougie being invariably followed by a severe rigor; further than this, the stricture was very contractile, and slow progress only was made. I considered this a suitable case for internal section. A fortnight after his admission, I was able to introduce Watson's urethrotome, with which I divided the stricture; afterwards I readily passed a No. 12 catheter and emptied the bladder. On the following day, instead of difficulty in micturition, he passed to the opposite condition, and complained of inability to hold his water. This I have noticed on several occasions; it is due to the suddenness of the relief afforded by the removal of the stricture, before the bladder can adapt itself to the altered conditions. In the course of a few days this new source of difficulty disappeared, as I had assured the patient would be the case, and the full size of the urethra was maintained by the introduction of a suitable bougie. This operation the patient was soon able to accomplish for himself, and when he left the hospital, which he did in a fortnight, all difficulty in passing water was removed.

I have recently been using, as some of you may have seen, a urethrotome which has been constructed for me by Messrs. Krohne & Sesemann, of London, and which is somewhat different from the instruments at present in use or figured in the text-books.

The staff consists of two parts—viz., the anterior portion, sufficiently small to pass into the narrowest strictures, and behind this an expanded portion, corresponding in size to a No. 10 bougie, and terminating anteriorly in an abrupt shoulder. Within the broad portion of the staff is contained a lancet-shaped knife, which is made to project by a spring at the handle, and run along a slit in the narrow part of the instrument. The extent to which the blade can be projected is regulated by a screw. It can be easily taken to pieces, for the purpose of cleaning or resharpening the knife, by unscrewing the stylet at the handle, when the blade becomes disengaged and can be taken from the slit. In the accompanying drawings the instrument is represented closed (Fig. 18), and with the knife projected (Fig. 19) as for the division of a stricture. When the instrument is passed down the urethra, the position of the stricture is indicated by the broad shoulder, against which it is made firmly to press, as in Syme's staff for perinæal section. The position of the stricture being thus ascertained and commanded by the broad shoulder, the knife is then made to project, and to divide the stricture. If the instrument is held firmly, the divi-

Fig. 18.   Fig. 19.

sion of the stricture is usually indicated by the feeling of tension giving way, something like when a tendon is divided for the cure of a deformity. The knife is then withdrawn, when, if the stricture has been completely divided, the broad shoulder passes on readily towards the bladder. Should any further obstruction be indicated, its division may in the same way be accomplished. I usually pass a full-sized catheter into the bladder, to make sure that all obstruction is completely removed and to draw off any water that may be there.

It will be seen that by this instrument the urethra is divided at the strictured spot in two places instead of one. I believe this will be found to be an advantage, inasmuch as contraction is less likely to follow where we have two longitudinal intervals in which healthy repair is allowed to take place. To maintain the interval made by the urethrotome until such time as cicatrization has been accomplished, I employ in the subsequent treatment of these cases oval-shaped bougies.

The incisions that are made by the urethrotome are as represented in this sketch. My object is to open up these incisions and maintain them patent, so that, when cicatrization is completed, the section of the urethra, if put on the stretch, would be thus *
To effect this, after division by the urethrotome, I employ oval-shaped bougies, sections of which are here represented, by means of which dilatation of

---

* Putting in what Dr. Gouley describes as two "cicatricial splices."

the urethra in a lateral direction is specially maintained:

In the largest sizes the circumference of the bougies gradually diminishes towards the handle, so that the meatus of the urethra is not kept continuously on the stretch whilst they are being introduced.

So far as I have practised this mode of treatment, I have every reason to be satisfied with the result; time, however, must elapse before any advantages it may possess can be properly estimated.

Whatever instrument we may select for the performance of internal urethrotomy, there are certain conditions or rules to be remembered which are essential to its safety and success. First: we must take care that complete division of the contracted part or parts of the urethra can be effected. I have seen internal urethrotomy resorted to in a case where the mass composing the stricture was so extensive as to be amenable solely to perinæal section. Under such circumstances failure could alone be anticipated. Second: the operation should not be performed where there is any active suppuration going on in the urethra. Apart from the risk of incurring septicæmia, it is desirable that the incision made should heal as kindly as possible, and without the intercurrence of that which interferes with healing, viz., active inflammatory action. Mr. Lund lays considerable stress on this point in his very interesting essay on internal urethrotomy.* Third: on the completion of the operation I leave a catheter in the bladder for forty-eight hours, connecting it with a piece of india-rubber tubing, by means of which the urine is constantly, and to the patient imperceptibly, running off into a receptacle placed by the side of the bed. This plan keeps the urethra outside the catheter much drier than the old way of letting the patient remove a plug from the orifice of the instrument whenever he feels a desire to micturate.

The risk of hæmorrhage has been advanced against internal urethrotomy. I may add, that where the operation is done with an instrument such as the one I am using and have described, no apprehension of this need be entertained, inasmuch as, instead of dividing the urethra pretty deeply in one place, the same effect is produced by a more limited incision at two points of the circumference. I do not commence the introduction of bougies until four or five days after the performance of the operation, and this I usually continue up to numbers 14 and 15, so as to make sure that the incisions are fully opened up.

In a recent paper, Mr. A. E. Durham † has advocated the performance of internal urethrotomy by means of an instrument which divides the stricture at four different points, as represented in this sketch — +. I consider that the use of such an instrument is very likely to be followed by slough-

* *On Internal Urethrotomy*, by E. Lund, 1877.
† *British Medical Journal*, March 18, 1878.

ing of one of the angles made by the urethrotome, and should that occur, the additional cicatrization that follows necessarily increases the amount of stricture material in the part operated on.

Before concluding this lecture, I will briefly notice a mode of treating strictures which, though differing materially from that which has just been described, has been very extensively practised. I allude to Holt's operation. The details of this procedure are so well known that I need not trouble you with them. The chief objections urged against it are, that the urethra is torn, frequently at its weakest spot in the strictured circumference, and that thereby the amount of cicatricial material is considerably increased. Hence, if the symptoms of stricture return, or rather are allowed to return, they are of greater severity and more difficult again to treat. There may be some amount of truth in these allegations, but I think the objections to the operation have been greatly exaggerated. I have practised it on something like one hundred cases. I never met with any difficulty in performing the operation; I have never had a fatal result; it has never been followed by serious hæmorrhage or by distressing symptoms, such as extravasation of urine; and though I cannot speak as to the ultimate results in anything like the majority of my cases, I know of many patients upon whom I operated eight and ten years ago who are living in complete comfort, and to all intents and purposes are rid of those urgent symptoms which necessitated this mode of treatment. Like many other surgical operations, it is frequently brought into discredit by disregard of those very conditions upon which its success depends.

Post-mortem examination has shown that two different results may be obtained by this operation. Mr. Christopher Heath mentions two cases operated upon in this way, where, though the strictures had been completely removed, the mucous membrane remained intact.* These undoubtedly were instances of submucous stricture, where the obstruction encircled the mucous membrane, just, as Benjamin Bell describes, as if a string had been tied around the urethra. At a meeting of the Lancashire and Cheshire Branch of the British Medical Association, held at Preston, in 1870, I showed the urethra of a patient upon whom Holt's operation had been performed shortly before death. I am indebted to Dr. Lyster for the specimen and particulars of the case. The operation was performed under urgent circumstances by Dr. Lyster, and it in no way appeared to contribute to the death of the patient, who was suffering from a dropsical affection. On opening the urethra, the stricture was found completely split, including the mucous membrane, commencing rather in front of the stricture, and extending backwards somewhat obliquely to one side.

These two sets of illustration show us what is actually done in Holt's operation. If we could limit it to the first class of cases—viz., submucous strictures—the operation would, in every respect, be a success, inasmuch as we should reserve, either for

* *British Medical Journal*, July 17, 1869.

dilatation or some form of urethrotomy, the cases where the mucous membrane is itself implicated. In practice we are able, to some extent, to give effect to this distinction, and when this can be done, cases of submucous stricture may advantageously be submitted to Holt's operation.

Where we have to do with more strictures than one, and the case is further complicated with false passages, or where the stricture is so irritable as to render gradual dilatation intolerable, I have frequently performed Holt's operation, in preference to any other mode of treatment, with the best results. Though internal urethrotomy has been brought to great perfection, and is now practised with precision, I am not prepared to admit that it should supplant an operation which, in properly selected cases, gives such good results as that generally known as the "immediate method."

NINTH LECTURE.

External Urethrotomy—Syme's Operation—Selection of Cases—External Urethrotomy with a Guide—Without a Guide—Wheelhouse's Operation—Subcutaneous Urethrotomy.

We have recently had under observation several cases where a section of the perinæum has been performed for the treatment of stricture or some of the complications arising from it.

You are accustomed to see in the wards of this Infirmary a large number of cases of stricture; for the most part these are remedied by a much simpler proceeding—viz., by gradual dilatation. Exceptional cases not unfrequently present themselves in which the usual mode of treatment has to be departed from, and more extreme measures resorted to. These exceptions to the rule will now occupy our attention.

For the division of stricture from without, as at present practised, we are largely indebted to the late Professor Syme. I do not mean to say that he originated this operation, but he taught us how it might be more efficiently performed, and to what class of cases it is applicable.[*] It may be generally stated that perinæal section is resorted to in the worst forms of stricture; and this includes—(*a*) strictures complicated with

---

[*] Described by the older French writers as "la boutonnière."

fistulous openings, through which the urine escapes, and where it is necessary to provide a free and direct vent, so as to bring about the closure of these openings. (*b*) Strictures which are not remediable by dilatation, such as the extremely contractile strictures which follow laceration of the urethra, and are too extensive for treatment by internal urethrotomy. (*c*) Strictures which, so far as a catheter or bougie is concerned, are impassable.

The application of perinæal section to these three classes of cases necessitates the performance of the operation under two very different circumstances. First, with a guide, where section of the stricture is certain and complete; and, second, without a guide. It is better to restrict the term "external urethrotomy" to the former, and "perinæal section" to the latter. When a staff can be passed into the bladder, the operation is comparatively simple. For its performance no better rules can be followed than those laid down by Syme. By carefully incising the perinæum from the raphé in the median line, the staff is reached, and by keeping the point of the knife in the groove, the stricture can be completely divided without injury to the surrounding parts. The knife must be kept undeviatingly in the central line of the perinæum; unless this rule is strictly adhered to, free hæmorrhage may be expected.

When such an operation has to be undertaken for impassable stricture, it is one of very considerable difficulty, often taxing to the uttermost the patience and perseverance of the surgeon. Under these circumstances, the most must be made of the two landmarks which such cases present; these are (*a*) the anterior extremity of the stricture, which is indicated by the point to which a staff can be passed, and (*b*) a dilated condition of the urethra, which is found behind strictures of long standing, and which, when the bladder is distended, is evident to the touch externally, or to the finger in the rectum.

In cases complicated with perinæal fistula through which urine has for a long time been passing, even the latter landmark—a dilated condition of the urethra behind the stricture—can hardly be said to exist, as the urine escaping through the fistulous passages fails to exercise that dilating power upon this part of the urethra which is usually observed in strictures that are not so complicated.

With these for landmarks, such as they are, to effect a junction between them, and so to reach the contracted urethra by division of the stricture, two plans of proceeding are practicable. One is by incising the perinæum along the median line down to the point at which the staff is arrested, and from thence backwards towards the dilated portion of the urethra. The other consists in reversing this order of proceeding by plunging the knife towards the dilated portion of the urethra behind the stricture, and then carrying the incision forwards toward the permeable part of the urethra. I have practised both plans with success, but I am disposed to think that the former method is the one more generally to be recommended.

When the operation by either method is successfully practised, the operator should be able to pass a catheter into the bladder, along the whole length of the urethra, so that he can satisfy himself that the stricture has been completely divided. I usually leave a catheter in the bladder, introduced through the perinæal wound, for twenty-four hours after the operation. One of Holt's winged catheters answers admirably for this purpose. Subsequently the treatment consists in the introduction of bougies along the urethra at intervals of four or five days, according to circumstances. Under this management the perinæal wound closes without trouble. The results of the operation have, so far as I have seen, been very satisfactory; I believe I have operated on over twenty cases without a death. The effect on the stricture has also been decidedly beneficial. I do not mean to say that the patient can hope to dispense with the occasional introduction of the bougie, but what he may expect is that, with a moderate amount of care on his own part, his stricture, which previously caused him all manner of distress, will by this operation be rendered easily manageable.

As I have already said, the operation of external urethrotomy without a guide is one of very considerable difficulty, and no one can undertake it without a misgiving that, though the proximal portion of the urethra may be opened up, the stricture may escape that complete division which is essential to the ultimate success of the operation. To meet these difficulties an operation has been practised by Mr. Wheelhouse, which is now known as the "Leeds operation."

Considerable credit is due to Mr. Wheelhouse for the thoroughly practical manner in which the operation has been planned, and, inasmuch as you will find only a meagre description of it in the text-books, I shall quote *in extenso* Mr. Wheelhouse's paper in the *British Medical Journal*, of June 24th, 1876 :—

"Notwithstanding the length of time that has elapsed since, in 1869-70, I brought before the profession, in the columns of the *British Medical Journal*, my method of finding my way, in cases of impermeable stricture from the perinæum, *through* the stricture and into the bladder, the subject seems to have received so little notice that I deem it advisable once more, after several years of successful employment of the operation, to revert to the subject; and I am induced to do this the more willingly, because I find that even in the most recent and most voluminous works on surgery, the subject is dismissed with very few words, and the old hap-hazard measure of reaching the bladder without any guide—it *may be through*, or it *may be altogether wide of, the stricture*—is still recommended and described as the one in ordinary use. Over this method, the procedure which I adopt has at least the advantage of greatly increased precision; it renders an operation, confessedly hitherto one of the most difficult in surgery, a comparatively easy one, and one which, in my hands and in those of my colleagues, has given results infinitely more favourable, both in immediate and ultimate effect upon our cases, than any we had ever seen before its introduction. The instruments required are as follows: lithotomy bandages; a special staff, fully grooved through the greater part, but not through the whole, of its extent, the last half inch of the groove being 'stopped,' and terminating in a round button-like end (Fig. 20); an ordinary scalpel; two

Fig. 20.

pairs of straight-bladed forceps, nibbed at the points; ordinary artery forceps and ligatures; sponge; a well-grooved and finely probe-pointed director; Teale's probe-gorget (Fig. 21); a straight probe-pointed bistoury; a short silver catheter (No. 10 or 11 gauge), with elastic tube attached.

"The patient is placed in lithotomy position, with the pelvis a little elevated, so as to permit the light to fall well upon it, and into the wound to be made. The staff is to be introduced with the groove looking towards the surface, and brought gently into contact with the stricture. It should not

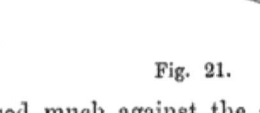

Fig. 21.

be pressed much against the stricture, for fear of tearing the tissues of the urethra, and causing it to leave the canal, which would mar the whole after-proceedings, which depend upon the urethra being opened *a quarter of an inch in front of the stricture*. Whilst an assistant holds the staff in this position, an incision is made into the perinæum, extending from opposite the point of reflection of the superficial perinæal fascia to the outer edge of the sphincter ani. The tissues of the perinæum are to be steadily divided until the urethra is reached. This is now to be opened *in the groove* of the staff, *not upon its point*, so as certainly to secure a quarter of an inch of healthy tube immediately in front of the stricture. As soon as the urethra is opened, and the groove in the staff fully exposed, the edges of the healthy urethra are to be seized on each side by the straight-bladed nibbed forceps, and held apart. The staff is then to be gently withdrawn until the button-point appears in the wound. It is then to be turned round, so that the groove may look to the pubes, and the button may be hooked into the upper angle of the opened urethra, which is then held stretched open at three points thus (Fig. 22), and the operator looks into it immediately

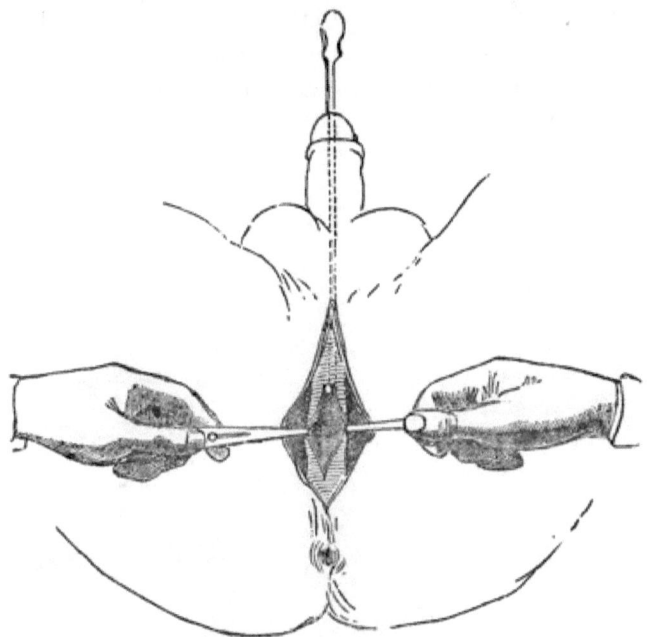

Fig. 22.—Staff introduced.

in front of the stricture. Whilst thus held open, the probe-pointed director is inserted into the urethra; and the operator, if he cannot see the opening of the stricture, which is often possible, generally succeeds in very quickly finding it, and passes the point onwards *through* the stricture towards the bladder. The stricture is sometimes hidden amongst a crop of granulations or warty growths, in the midst of which the probe-point easily finds the true passage. This director having been passed on into the bladder (its entrance into which is clearly demonstrated by the freedom of its movements), its groove is turned *downwards*, the whole length of the stricture is carefully and deliberately divided on its under surface, and the passage is thus cleared. The director is still held in the same position, and the straight probe-pointed bistoury is run along the groove, to ensure complete division of all bands or other obstructions. These being thoroughly cleared, the old difficulty of directing the point of a catheter through the divided stricture and onwards into the bladder is to be overcome. To effect this the point of

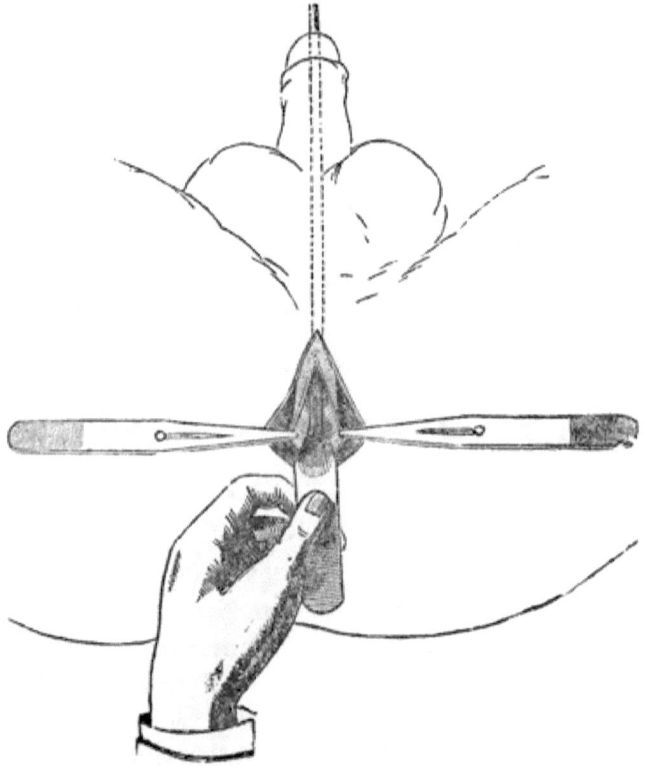

Fig. 23.

the probe-gorget is introduced into the groove in the director, and, guided by it, is passed onwards into the bladder, dilating the divided stricture, and forming a metallic floor, along which the point of the catheter cannot fail to pass securely into the bladder. The entry of the gorget into the latter viscus is signalised by an immediate gush of urine along it.

"The short catheter is now passed from the meatus down into the wound; is made to pass once or twice through the divided urethra, where it can be seen in the wound, to render certain the fact that no obstructing bands have been undivided; and is then, guided by the probe-dilator, passed easily and certainly along the posterior part of the urethra into the bladder thus (Fig. 23).

"The gorget is now withdrawn; the catheter fastened in the urethra, and allowed to remain for three or four days; the elastic tube conveying the urine away to a vessel under or by the side of the bed.

"After three or four days, the catheter is removed, and is then passed daily, or every second or third day, according to circumstances, until the wound in the perinæum is healed; and, after the parts have become consolidated, it requires, of course, to be passed still from time to time to prevent recontraction."[*]

Before concluding these observations, let me say a few words in reference to the application of subcutaneous surgery to the treatment of stricture. Under favourable circumstances a stricture may be divided from without subcutaneously. I have practised this treatment in a few instances with excellent results. It is limited, however, to penile stricture, where division can be effected without opening into the urethra, the obstruction being around the urethra rather than in it.

In a case of penile stricture in which I removed the urethra from a patient in the Northern Hospital, who

---

[*] I am very much indebted to Mr. Wheelhouse, for allowing me thus to make use of not only his paper, but also of his original woodcuts explanatory of the operation.

died of acute rheumatism, I found, on dissecting up the mucous membrane, that its dimensions were in no way altered, the obstruction being entirely around it. Such a case would have been suitable for subcutaneous division.

I think the term "subcutaneous," as applied to the mode of operating described by Mr. Teevan,* or that by Dr. Dick,† inappropriate, as the conditions are dissimilar to those we recognize in other forms of subcutaneous surgery; for instance, in tenotomy. In the description just referred to, the stricture is divided by a punctured incision through the perinæum, on a grooved staff, or a modification of it, passed through the stricture. Though the wound may be subcutaneous so far as the skin is concerned, yet inasmuch as the urethra is laid open, the section is exposed to the irritating influence of the urine flowing over it, or of the catheter that may (as in Mr. Teevan's case) be retained; hence it appears to me that such proceedings have no further advantage than those which belong to a properly performed internal urethrotomy.

As bearing upon the respective merits of two operations—namely, dividing strictures from within or without the urethra—the communications are of much value, and will be read with interest. I, however, share with Mr. MacCormac the opinion he seems to have expressed at the discussion following Mr. Teevan's paper, that the term "subcutaneous" was

---

* *Subcutaneous Urethrotomy*, by Mr. Teevan; Clinical Society's Transactions, vol. viii.

† *Subcutaneous Division of Stricture*, by H. Dick, M.D.

hardly applicable, however well devised the operation might be.*

For the reason that the operations referred to do not appear to me to present any advantages over an internal urethrotomy, it will be unnecessary further to allude to them.

* *The Lancet,* Jan. 16, 1875, p. 89.

TENTH LECTURE.

Syphilitic Strictures—Nunn's and Bell's Views—Chancre of the Meatus—Cases of Stricture Complicated with Syphilis—Treatment.

In a paper read before the Medico-Chirurgical Society, in 1866, Mr. Nunn, of the Middlesex Hospital, pointed out that many cases of stricture were of a syphilitic nature, and required for their treatment the employment of those means which exercise a curative power over this specific disorder.*

Though the pathology of venereal disorders was not then worked out as it is now, such a shrewd practitioner as Bell was not likely to be deceived in a matter of clinical observation. I have on several occasions had the opportunity of verifying the correctness of these views, and of recognising their value in the treatment of what, at first sight, appeared to be the worst forms of stricture, and I therefore consider it

---

* Mr. Nunn has kindly referred me to a passage in Benjamin Bell's *System of Surgery*, vol. 2, p. 221, fifth edition, where he states that "whatever may, in disorders of this kind, be the immediate cause of obstruction to the free passage of the urine, a venereal taint will for the most part be found to be the original cause of the whole. We have therefore desired that, at the same time the use of bougies is persisted in, the patient ought to be put upon a very complete course of mercury, in order to destroy every possibility of his suffering again from the same cause; for we need scarcely observe, that as long as any venereal affection continues to prevail, little or no permanent advantage can be expected from the use of bougies or any other remedy."

desirable that I should again draw attention to this complication.

I do not refer to the puckerings consequent on the healing of venereal ulcers, which are usually met with at the meatus, but to the deposition, about the urethra, perinæum, or scrotum, of the syphilitic exudation, which, becoming organised, often in considerable masses, impedes the dilatation of the canal. At first sight, from their extent and their extreme induration, these strictures present a most unfavourable aspect, but on diagnosing their nature, they eventually prove the most satisfactory to treat; thus again pointing us to the importance of recognising the constitutional origin of local disease.

I shall now proceed to illustrate certain forms of stricture having a syphilitic character, and to remedy which anti-syphilitic treatment must be resorted to.

And, in the first place, I would observe that an ordinary chancre at the meatus may act as a stricture, and be mistaken for one.

Some time ago I was consulted by a gentleman for what he was led to believe was a stricture of the meatus. His history was, that some three months previously he had a chancroid sore on his penis, which in the average time was healed by the usual remedies. The sore was considered to be of a local nature, and no secondary consequences were anticipated. Just about the time this sore was healing, an induration appeared at the meatus of the urethra, accompanied by a slight gleety discharge. This slowly but steadily increased until it involved about the last inch of the urethra.

As the induration increased, so did the difficulty in making water.

Upon examination, I observed the scar of the first sore in the sulcus, which was soft and apparently non-syphilitic. The orifice of the urethra was almost completely occluded by an indurated mass extending downwards for nearly an inch, and inverting the lips of the canal. I could pass through it with some difficulty a probe, and on squeezing the urethra a small quantity of thin watery discharge exuded. One or two glands in the groin were indurated, but not extensively. Though in some respects this case had the appearance of an ordinary stricture at the meatus, there could be no difficulty in arriving at the conclusion that the obstruction was caused by a chancre. I advised the discontinuance of local treatment by bougies, and that the patient should be put under the influence of mercury. In eight weeks the induration had almost entirely gone, and with it all difficulty in passing water.

The obscurity of diagnosis was in the sequence of the two forms of venereal sore—viz., the locally contagious and the infectious. I believe this to be an instance of double inoculation occurring at one time.

The next case illustrates better the connection which occasionally exists between stricture and syphilis.

In 1875, a gentleman consulted me for a stricture, which he informed me had been caused by an accident in the hunting field. From the description of the injury, it was clear that if his urethra had not been actually ruptured, it had been contused; I think more probably the latter. At the time of the accident he was suffering from undoubted secondary syphilis.

For some weeks after the injury he experienced little or no inconvenience in passing water; he continued to take horse exercise, and paid no attention to the constitutional symptoms from which he was suffering.

In the course of three months after the injury, symptoms of stricture appeared, and progressed rapidly, accompanied with induration behind the scrotum.

When I first saw him the signs of stricture had extended over seven months. The scar of the syphilitic sore was still visible on the penis, and remained indurated. Two or three small glands in the groin were also enlarged and similarly affected. There was considerable induration to be felt in the perinæum, immediately behind the scrotum, and more so than appeared explicable by the injury he had received. I was consequently led to enquire closely into his history, with the result I have mentioned.

Upon examining the patient with a bougie, I found that a No. 3 was grasped with considerable tightness.

I came to the conclusion that the case was one of traumatic stricture, complicated with syphilis. I placed the patient under the influence of mercury, and commenced treatment by gradual dilatation. In the course of three months the induration in the perinæum, that was at first palpable to the touch externally, entirely disappeared, and a full-sized instrument could be readily passed. I should also state that the hardness alluded to in the cicatrix on the penis and in the glands also subsided under the influence of the mercury. The patient has since continued to pass a bougie for himself, and has had no further difficulty.

The third and last case I shall mention was under treatment in the Infirmary, where I took the opportunity of drawing attention to the special complication. The patient had suffered from stricture, which he attributed to a gonorrhœa. For two years nearly the whole of the urine had been freely passed through a fistula behind the scrotum, but little escaping by the natural channel. The perinæum, scrotum, and the edges of the fistula were as hard as cartilage, a condition which I at first attributed to the irritating influence of the urine and the persistence of chronic inflammation about the parts. In this case I also found that the patient had suffered from syphilis about three years previously, and there still remained further evidence of the disease. The throat was deeply scarred, there was an induration in the centre of the tongue, and an indistinct thickening of the periosteum on one clavicle, and over the shin. The patient was a sailor, and, according to his statement, had never received any treatment for his syphilis. With this history before me, and seeing that no serious difficulty was experienced by the patient in passing his water, I resolved to rest content for the present with placing him under the influence of mercury. This was done and maintained for several weeks. Under treatment the patient improved in a remarkable manner, and the indurations disappeared to a very considerable extent. In eight weeks I commenced regular treatment by bougies, and before the patient left the hospital I had the satisfaction of finding the fistulous opening close, and the urethra capable of receiving a full-sized bougie.

I might further exemplify the connection that frequently exists between syphilis and stricture, but I think the cases I have recorded are sufficient for this purpose.

My reference to syphilis has for its special object the reminding you of its occasional occurrence as a complication of stricture, under such circumstances as I have endeavoured to illustrate, and my remarks as to special treatment will be exceedingly limited.

Where the syphilitic taint exists, excepting in the case of some forms of its tertiary stage, the patient should be brought under the influence of mercury, which is to be maintained until all traces of the infection, constitutional or local, have ceased to exist. There is no other drug upon which reliance can be placed. The outcry that one occasionally hears in reference to mercury is a relic of bygone days, when its abuse was thought to be its use, clinical observation then falling far short of the precision which now obtains.

Some most important investigations have recently been made bearing upon the tonic influence that mercury is capable of exercising.

Most of us must frequently have observed how persons improve in condition, gain weight, and become sleek, under its action.

Dr. Keyes, of New York, has paid considerable attention to this subject, and demonstrated, I consider, in a most conclusive manner, "that mercury in minute doses is tonic in all cases where it can be digested, in syphilis, or out of it, continued for a short or a long (over three years) time."

By some carefully-conducted observations, and the use of the hématemètre, he has shown,—

"That syphilis diminishes the number of red corpuscles below the healthy standard."

"Mercury in small doses, continued for a short or long period in syphilis, alone or with the iodide of potassium, increases the number of the red corpuscles in the blood, and maintains a high standard of the same." *

Thus is explained that which clinically is a matter of observation.

Such, then, are the effects of mercury we wish to obtain. Where syphilis occurs as a complication of stricture, I prefer commencing the mercurial treatment by a course of inunction. It is, I know, somewhat old-fashioned, but experience shows us that it has many advantages, one being the readiness with which you can control its effect on the system. Do not suppose that I recommend inunction in the manner that appears to have been practised so late as the time of Sir Astley Cooper, who, in denouncing its abuse, speaks of its "exhibition," as the old phrase goes, producing "three pints of saliva a day." † No wonder that the remnants of prejudice still exist.

---

* *The Tonic Treatment of Syphilis*, by E. L. Keyes, M.D. New York, 1877.

† *Lectures on Surgery*, 8th ed., p. 448.

Inunction may be employed either by means of the mercurial ointment rubbed into the thighs, or the oleates of mercury, which are convenient and suitable, the amount of mercury in them varying, as may be required, from five to twenty per cent.

Anything like an excessive action should be avoided, the object being to accomplish what is required with as little disturbance to the patient as possible. Salivation is under all circumstances to be deprecated.

If any further evidence of the value of inunction in the treatment of syphilis were necessary, I could refer to the extended experience of Dr. Brandis, of Aix-la-Chapelle, under whose advice it is so largely employed in connection with the natural resources of this fashionable retreat.

When inunction is impracticable, either from the patient being unable to resort to it, or from the effect produced by the mercurial ointment on the part to which it is applied, I prefer the perchloride or the proto-iodide of mercury, the latter being advantageously combined in a pill, with a little Dover's powder to prevent purging.*

Local treatment, usually by dilatation, will also be necessary, and it is astonishing to note how easily this proceeds as soon as the system becomes conscious of the presence of mercury. I have tried smearing the bougie used with mercurial ointment, but without advantage; the amount so introduced is not sufficient

---

* An excellent paper containing an account of the various modes of administering mercury, by Dr. J. Duncan, appears in the *Edinburgh Medical Journal*, 1877.

to be of any avail, whilst it is sometimes found to irritate the urethra. For the same reason, I do not employ urethral pessaries, preferring to produce the general effects of the mercurial, which really is all that is required, by one or other of the plans usually adopted for this purpose.

In these cases, whenever the obstruction is relieved and the urgency of the distress is over, let the patient be induced for his own safety to continue anti-syphilitic treatment until all signs of the constitutional disorder have disappeared. There is nothing to be feared from the judicious employment of mercury for the time that is necessary to destroy the effects of syphilis, whilst there is everything to be apprehended from a disease which, if uncontrolled, is sure to follow its own morbid inclinations.

ELEVENTH LECTURE.

CONSEQUENCES OF STRICTURE — URETHRAL ABSCESS — EXTRAVASATION OF URINE.

I HAVE already indicated generally the consequences which stricture may occasion within the urethra and beyond it. To certain of these complications I purpose to make a further allusion.

A not unfrequent consequence of long-standing stricture is the formation of an abscess in immediate connection with the urethra; and where this takes place a serious aggravation of the mischief usually ensues, exposing the patient to the risk of extravasation of urine, fistula, and impervious urethra. To explain the formation of abscess, we have only to notice the changes that take place in the urethra immediately behind the stricture. These have been observed to be a dilatation and thinning of the walls of the canal and of the lacunæ and ducts opening into it. These changes are chiefly to be seen in the floor of the urethra, being caused by the pressure exercised by the bladder to force the urine through the contraction. In this way the urethra becomes a receptacle for urine, which, undergoing decomposition, sets up inflammation and suppuration both within and around it. Abscesses so formed may, by ulceration, open into the urethra, and cause extravasation of urine. It is my belief, as I have already stated, that extravasation of urine as a consequence of stricture commonly occurs in this way, and not from rupture of the urethra by the expulsive efforts of the patient; or, the abscess may result in the formation of a fistula, through which the urine, in part, or even entirely, is passed. Such consequences as these may usually be averted by prompt treatment. No rule in surgery is more uncompromising than that which requires the use of the knife in the case of all acute inflammatory swellings in any relation with the urethra. The neglect of this rule, for which in many instances the patient is solely responsible, has led to a disastrous destruction of tissue, and extreme danger to life. The risk of an urinary fistula is little to be compared with the damage that extravasated urine may inflict, and should never be taken into consideration.

In incising the perinæum to relieve an abscess, I have sometimes included the division of the stricture where I have been able to pass a guide into the bladder, as in the following case, which is a good illustration :—

J. B., seaman, aged 30, was admitted into the Infirmary under my care on January 11th, 1870. He has had stricture for nine years. Recently he has been suffering from extreme irritability, his nights especially being much disturbed. The stream of urine has been gradually diminishing in size.

On admission into the Infirmary, the bladder was found distended, the urine issuing incontinently in drops. There was a perinæal swelling, with an indistinct sensation of fluctuation.

The stricture was at the membranous portion of the urethra. The patient had had a rigor two days previously. As extravasation of urine was imminent, I had the patient placed in the lithotomy position, and, putting my index finger in the rectum, I introduced a long straight finger knife into the median raphé an inch in front of the anus, and from the apex of the prostate gland divided in a direction forwards all the indurated and infiltrated tissues of the perinæum. This gave vent to some pus and a considerable gush of urine. I was then enabled to pass a centrally grooved staff into the bladder, and to complete the division of the stricture. There was no bleeding to speak of. I retained an elastic catheter in the bladder for some hours, and then removed it, allowing the urine to escape both by the urethra and through the perinæum. The after-treatment consisted in the introduction of bougies at regular intervals. The patient left the Infirmary on February 1st with the perinæal wound completely closed, and a No. 12 bougie passing readily.

As I have already stated, urinary abscesses are not unfrequently followed by fistulous openings; I shall, however, reserve the treatment of urinary fistula for a future occasion, as the subject is one of very considerable importance.

When urine escapes from its normal channel and becomes diffused amongst the tissues, extravasation or infiltration is said to have happened. The effect produced upon the tissues with which urine comes in contact is to destroy them; hence it is of the first importance to give free vent, with as little delay as possible, to that which has been extravasated.

Though in no way altering the rules for our guidance in the treatment of extravasated urine, I may mention that Menzel* has demonstrated that this destructive effect on the tissues is the property alone of decomposed ammoniacal urine, and does not belong to that which is limpid and healthy, the latter, when effused, being capable of absorption.

That urine may occasionally be extravasated without causing destruction of the tissues with which it is in contact, is referred to in connection with a case narrated in my lecture on the examination of the urine, where there was a most marked absence of urea, leading to the inference that the destructive power of retained urine on the tissues is through the generation of ammonia by the decomposition of its urea.

In practice, however, when urine is extravasated, we draw no such fine distinctions, it being the duty of the surgeon to provide channels for its escape, and so prevent that inevitable destruction of tissue which, almost without an exception, follows.

Extravasation may occur in connection either with urinary abscess or with laceration of the urethra. The amount of skin which perishes is sometimes very extensive. In the following case the whole of the scrotum was in this way destroyed:—

J. M——, aged 30, was admitted into the Northern Hospital under my care on November 4th, 1867. For several months he has experienced difficulty in passing water, and for the last week he has only voided it in drops. Three days before admission, after violent straining, his scrotum began to swell, and increased in size to an alarming extent.

* *Wien. Medizin. Wochenschrift*, No. 81–85, 1869.

On admission to the hospital the whole of the scrotum was gangrenous, and the lower part of the abdomen œdematous. The perinæum was not swollen. A stricture existed four inches from the meatus. The cavity of the right tunica vaginalis was full of fluid.

Free incisions through the sloughy mass were made. The hydrocele also was punctured. The whole of the scrotum came away, leaving the testicles completely exposed and pendulous. On November 10th there was a discharge of pus from the urethra. Gradual dilatation was subsequently employed. After a very tedious convalescence the patient left the hospital quite well. His testicles, however, were not very comfortably located, the new scrotal receptacle for them formed by granulation being inconveniently limited.

The direction extravasated urine takes is determined by the connection of the fasciæ in relation to it. This is explained by a reference to Fig. 6.

When extravasation takes place behind the triangular ligament, it is almost invariably fatal by the induction of pelvic cellulitis. We see this in cases of lithotomy in which the knife has been used with too great freedom on entering the bladder. When urine is conducted beneath the skin of the abdomen, it may travel as high as the umbilicus, as in the following case, where so much of the abdominal parietes was destroyed as to interfere with micturition :—

Charles W——, aged 30, was admitted into the Infirmary under my care on September 30th, 1875.

The patient, when about ten years old, fell on an iron rail, injuring his perinæum. A stricture followed and was treated in the usual manner, but ever since he has suffered from severe attacks of retention, requiring catheterism. About five years ago he had an attack of gonorrhœa, since which his stricture has been worse and retention of urine more frequent. A week before admission, he had increased difficulty, passing his urine in drops and with much pain. On September 30th, in violently straining to make water, he felt something give way, accompanied with an immediate sensation of relief. This was followed by rapid swelling of the scrotum, extending to the abdomen. He was then removed to the Infirmary. The perinæum and scrotum were enormously swollen, the swelling extending upwards to the abdomen. The bladder, apparently, did not contain much water. The patient was placed under the influence of ether, when the perinæum was freely incised, the urethra being opened on a staff so as to ensure the complete division of the stricture. Free incisions were also made in several directions into the scrotum. A No. 8 elastic bougie was put into the bladder through the urethra, and retained. The urine continued to escape, partly through the catheter and some through the wound. On the following day it was found necessary to make further incisions on either side of the abdomen, the urine having passed upwards almost to a level with the umbilicus. At the expiration of forty-eight hours the catheter was removed, but it was found necessary to replace it, the patient being unable to pass urine without it. There was no difficulty in getting a full-sized catheter into the bladder, so that I apprehended the inability to pass water was due partly to the previous over-distension of the bladder, and partly to the damaged condition of the abdominal muscles, which rendered any voluntary propulsive efforts on the part of the patient exceedingly painful. It was not until the abdominal wounds began to fill up with granulation that the patient recovered the power of micturition. A considerable portion of the scrotum mortified, and large sloughs of cellular tissue were discharged through the incisions in the abdominal walls.

The patient made a good recovery, and left the Infirmary on November 22nd. The wound in the perinæum and the other incisions closed up. At the time of his leaving the Infirmary a full-sized bougie passed readily.

There is one sign in connection with extravasation which is usually regarded as fatal. Brodie states:* "Sometimes a black spot may be seen on the glans penis; it is a most fatal sign, for I never knew one to recover in whom it appeared. It indicates that the urine has been effused into the cells of the corpus spongiosum." I have seen two examples of this, in each instance with a fatal result.

In the treatment of extravasation there are two points which must be given effect to: first, to prevent any further infiltration occurring; and, second, to secure a means of escape for the urine which has already become diffused amongst the tissues. The first indication is met by providing a direct escape for the urine by a perinæal or other suitable incision, and the second by further incisions, wherever there are grounds for believing that urine is lodged.

In cases of acute extravasation, the scrotum is usually enormously swollen and tense, and very shortly shows signs that mortification is impending. After incising the scrotum sufficiently to permit of the escape of urine from this part, in addition to other incisions that may be required, I am in the habit of squeezing such parts as the scrotum with my hand, just as I would do a wet sponge. It is astonishing what a quantity of urine, which would decompose and

* Brodie on the *Diseases of the Urinary Organs*, p. 14.

add to the sloughing, may thus be got rid of. In this way I have in a few minutes brought down an enormously distended scrotum nearly to its natural dimensions, and thus materially lessened the extent of tissue-destruction.

Extravasation of urine, when it occurs in young children, is almost invariably due to the impaction of a calculus in the urethra; I do not think I ever saw it happen otherwise but in one case, where the extravasation was due to an extremely contracted phimosis. Impacted calculi have in this way caused very extreme damage, the urine spreading into the cellular tissue of the scrotum and abdomen, and producing the same effects as occur in the adult. The principles of treatment are the same: remove the obstruction in the urethra, and make artificial channels for such urine as may have passed amongst the tissues.

In the case of an adult I recently examined in consultation at the Bootle Borough Hospital, one of the largest calculi I ever saw of the kind had made its escape in this manner. The patient was admitted under the care of Dr. Wills, with a sloughing scrotum and extensive extravasation of urine. The latter was relieved by incision, and from the scrotum, amidst a sloughy mass, a phosphatic calculus, weighing eighty grains, and moulded to the shape of the prostatic portion of the urethra, was removed. This, undoubtedly, had been the cause of all the trouble. The patient made a good recovery. An examination of a section of this calculus shows that it must have been in the urethra for a considerable time before it gave rise to serious symptoms of obstruction, as the wave-like lines of the formation indicated the probable increase of the stone by successive phosphatic depositions as the urine flowed out.

In the future management of these cases of extravasation it is most important that the strictest cleanliness should be insisted upon. Where there is much sloughing the wounds are necessarily offensive, and great care is required to keep them sweet. They should be syringed out regularly with tepid carbolic lotion, of the strength of one part of acid to sixty of water. In using carbolic acid in the local treatment of urinary affections, it is well not to have the solutions, be they oil or water, too strong, or they may occasion discolouration of the urine, which, though unattended, so far as I have seen, with any ill effects to the patient, is apt, under these circumstances more especially, to create some alarm.

Charcoal poultices are the best application during the separation of the sloughs, and, after this, water dressing, or some slightly stimulating lotion.

In all these cases I have the patient well packed between the thighs and elsewhere with picked oakum, such as is generally used in hospitals. It assists in disinfecting, and the tarry smell is not a disagreeable contrast to that which is exhaled by wounds and suppurations of an urinous nature.

Where there has been much destruction of the scrotum, advantage may be gained, after the sloughs have separated. by drawing the parts together with strapping.

In insisting upon the strictest attention to general hygienic measures, it must be remembered that septicæmia and pyæmia are the more frequent causes of death after these operations. That sickly urinous smell which so often hangs about the apartment and appurtenances of the urinary invalid is not uncommonly due to a want of cleanliness on the part of the patient or his attendant.* The ventilation of the apartment must be carefully seen to; remembering that cold, by checking the action of the skin, invariably increases the distress of the patient suffering from an urinary complaint. Where there has been much damage done by extravasated urine, the patients are apt to pass into a low typhoid state; and when the pulse and dry state of the tongue indicate this, stimulants and highly nutritious fluid foods are required. In one or two of the cases recorded the danger from this condition was great. The nervous irritability which is so frequently observed in these cases I have found best relieved by small doses of morphia hypodermically administered.

My remarks hitherto have had reference to abscesses connected with stricture.

Abscess may occur independently of a stricture in any of the tissues in relation with the urethra.

It may be caused by such irritants as the gonorrhœal discharge, improper injections, or the passage of instruments along the urethra. I attended a case with

---

* In keeping the bed of the patient sweet I have found a very simple contrivance, which was first suggested to me by Mr. Long, exceedingly useful. It consists in placing in the patient's bed a perforated wooden pill-box, containing a dozen grains or so of pure iodine

Mr. Swindon, of Wavertree, where there could be no doubt that bicycle riding was the cause.

When there are signs of local inflammation, means should be taken, by the employment of leeches, fomentations, and other emollients, to procure resolution, but there must be no hesitation in giving escape to matter should there be indications of its formation. Unless this is promptly done, it may burrow into the urethra and great damage result.

I have already stated, in my lecture on the anatomical relations of the urethra, that inflammatory exudation, when it occurs round the membranous portion of the urethra and between the layers of the fascia in connection with it, sometimes gives rise to the impression that abscess of the prostate is impending. I believe that a suppurating prostate is an affection far more rare than the text-books would lead us to believe. A careful examination by the rectum, with the index finger feeling the prostate, and the thumb of the same hand on the perinæum, is the best way of determining the position of inflammatory exudation in this part, and so of avoiding the error I have referred to.

Where there is no stricture, and the exudation around the urethra is not sufficiently pressing to cause retention of urine, the surgeon's endeavours to procure resolution may with safety be prolonged until the existence of pus is distinctly indicated by the sense of fluctuation. The occurrence of a rigor may not indicate much; I have often observed that a very slight interference with the act of micturition is sufficient in some persons to produce this symptom.

Where there is a stricture with retention of urine, it is proper rather to anticipate the formation of matter in an inflammatory swelling in relation with the urethra, inasmuch as the propulsive efforts of the patient, added to the tension already thrown on the urethra by the inflammatory peri-urethral exudation, may result in that rapid ulceration of its walls which so often precedes extravasation of urine. The distinction I have thus endeavoured to draw has not unfrequently enabled me to depart from that rule in surgery which renders imperative the early incising of inflammations in proximity with the urethra.

In opening an abscess in the perinæum, the result of gonorrhœa or such-like irritations, where there is no stricture causing retention, care must be taken not to open the urethra, as this procedure introduces an unnecessary element of danger, from which a simple perinæal incision is free; for even if, under these circumstances, the abscess should by ulceration make its way into the urethra, all risk of extravasation is avoided by the opening which has been made.

TWELFTH LECTURE.

INJURIES TO THE URETHRA — CONTUSION — RUPTURE OF THE URETHRA — CASES — TREATMENT — LONGITUDINAL WOUNDS OF THE URETHRA.

INJURIES to the urethra not only expose the patient to the risk of retention and extravasation of urine, but are usually followed by a stricture of the worst form.

In a seaport like Liverpool a very considerable proportion of the stricture cases are met with amongst its sailor population, and of these, a large number I have observed are traceable to injuries received in the course of their employment. Such, for instance, as blows on the perinæum by falls over ropes and spars, where the urethra is violently stretched or contused against the pubic arch. A contusion of the perinæum, without any obvious tearing of tissue, is quite sufficient to occasion retention, requiring the use of the catheter; but inasmuch as the urethra escapes without laceration, the inconvenience is a temporary one, and the formation of a stricture need not be apprehended.

There was a patient not long ago in one of my wards who was injured in this way. He had fallen from a considerable height over some railings, alighting upon his perinæum. On admission into the Infirmary some hours after, the perinæum and scrotum were very much swollen and ecchymosed. Though he could not pass water, yet there was no difficulty in introducing a catheter into the bladder. Nothing like a rent could be felt, and no blood appeared to have escaped from the orifice of the urethra. It was a case of contusion with very considerable ecchymosis, and after the lapse of forty-eight hours no further catheterism was required. Had there been evidence that the urethra was torn, it might have been necessary, as I shall presently state, to adopt a very different treatment. Actual laceration of the walls of the urethra may be caused from within, as in attempts to pass catheters, or in the expulsion or removal of calculi. Occasionally such an injury is inflicted by a fractured pubic bone, as in the case to which I shall presently refer.

It is, however, to the consideration of lacerations caused by violence applied externally that I wish now to direct attention, and with a view of making observations on the treatment of these injuries, I purpose to narrate the following cases which I have collected.

CASE 1.—A labourer, aged 20, was admitted into the Northern Hospital under my care in August, 1866. Eighteen hours before admission he had received a kick on the perinæum. Blood issued from the orifice of the urethra, and he shortly afterwards found himself unable to pass water. For this he applied at the hospital. With some difficulty a No. 7 catheter was introduced, a distinct laceration being felt about the bulb. Two pints of bloody urine were removed, and the catheter was then retained in the bladder. No further difficulty was experienced. The patient made a good recovery. A contraction about the lacerated part ensued, and for several months the patient attended as an out-patient for the purpose of having the urethra dilated by bougies. Eventually he was lost sight of, some contraction then remaining.

CASE 2.—A dock labourer was admitted into the Northern Hospital under my care in 1866, having fallen from a height astride over some scaffolding.

On admission the perinæum was bruised, and blood was passed by the urethra. On introducing a catheter, a rupture of the urethra, about the triangular ligament, was made out. The laceration did not appear completely to sever the urethra, for with a little trouble the catheter was introduced into the bladder. In this position it was retained, and up to the fifth day the patient appeared to make satisfactory progress. On this day, however, the patient became feverish; the perinæum was found swollen, and there was much pain about the part. Under these circumstances I opened the perinæum freely, dividing the urethra forwards from the apex of the prostate. Vent was thus given to disorganised clots and some fetid pus, and a certain amount of immediate relief was afforded. On the ninth day from the injury there was a rigor, and the patient rapidly succumbed, with well-marked symptoms of pyæmia. I was only able to inspect the injured part after death, when I found that the urethra had been almost completely, though irregularly, ruptured in the membranous portion. There were also signs of rapidly extending pelvic cellulitis.

CASE 3.—About the same time as the previous case, a sailor was under my care at the Northern Hospital for very similar injuries caused by falling across a rope.

Here the signs of a rupture of the urethra in its deeper part were equally unequivocal. With some difficulty I introduced a grooved staff, and freely laid open the perinæum and urethra to an extent sufficient to secure a free vent for the urine.

The patient made a good recovery, and the perinæal wound completely healed. During the healing process bougies were regularly introduced. I saw the patient not long ago, and he appeared to suffer no inconvenience from his accident. The urine was voided in a natural stream.

CASE 4.—A 'bus conductor, aged 17, was admitted into the Infirmary under my care in 1869, with the following history. Seven days before admission he was kicked on the perinæum; this was followed by slight hæmorrhage from the urethra, but he was able to pass his water.

On the day of his admission to the Infirmary (the seventh from the accident) the hæmorrhage recurred, and he had retention of urine. A No. 9 catheter was introduced, and a considerable quantity of urine withdrawn. Two days after this he became feverish, the perinæum was swollen, and there was some pain about the part. To relieve this a free incision was made into the urethra along the central raphé, through which all urine passed. This was followed by relief. He made a good recovery, and left the Infirmary with the perinæal wound completely closed. The urethra appeared to be lacerated about the membranous portion along its lower wall. During the process of healing, bougies were introduced in increasing sizes. He left the hospital with a urethra admitting a full-sized bougie, and I never heard afterwards that he suffered from stricture.

CASE 5.—A stonemason was admitted into the Infirmary under my care in 1870. He had fallen across a sharp stone, considerably bruising his perinæum.

On his admission he had the usual symptoms of lacerated urethra—viz., blood issuing from the orifice of the urethra, and retention of urine. A catheter disclosed a considerable rupture about the deep perinæal fascia. I accordingly laid open the perinæum freely in the middle line, giving exit to clots and forming a passage for the urine to escape.

The patient had a rather sharp attack of orchitis, but with this exception made an excellent recovery. The treatment consisted in the regular introduction of bougies whilst the perinæal wound was healing. When he left the Infirmary the wound was closed, and the urethra of its natural size. I saw him some months afterwards, and he had remained quite well without any sign of stricture.

In reference to the question of treatment, the cases I have recorded justify a conclusion that in all such injuries external incision, or, as it is more commonly called, perinæal section, is the safest plan of proceeding, recommending itself on the following grounds:—

1st. Because of the impossibility of accurately determining the extent and direction of the laceration.

2nd. Because incision is the surest means of preventing extravasation of urine; and

3rd. Because incision diminishes the risk of a stricture forming, or, at all events, moderates the severity of such a formation.

The relative position of a laceration to the deep perinæal fascia is a matter of the first importance. Were it possible in all cases to ascertain that the lacerated portion was anterior to the deep fascia, provided a catheter could be introduced into the bladder, it would be safe to treat the injury without incision, resorting to such a proceeding should signs of extravasation of urine appear; for, under these circumstances, the direction taken by extravasated urine is forward towards the scrotum, where it renders itself unmistakably apparent from the moment of its occurrence. Where the laceration is behind the deep fascia, the extravasation, should it follow either immediately or in the course of a few days, is of a much more serious nature, inasmuch as the urine takes a backward direction towards the pelvis, setting up cellulitis, which speedily goes on to suppuration. Here it is much more subtle; it may be going on from the moment of the injury, not declaring itself until it has occasioned symptoms of pelvic cellulitis.

The former variety of extravasation is usually amenable to treatment, but the latter is most frequently followed by a fatal result. Case 2 illustrates this, and Case 4, though the patient was saved, undoubtedly is a similar example. Inasmuch, then, as the precise position of the laceration, whether it be a few lines in front of the deep fascia or a few lines behind it, determines in a great measure the after-consequences, so far as extravasation of urine is concerned, is it not better to act on the safer side in these cases of deep laceration, and anticipate the risk of retrograde inflammation?

The second proposition, that incision is the safest means of preventing extravasation, naturally follows on admitting the impossibility of precisely determining the position of the injury to the deep fascia. No other plan than that of opening up the injured spot, and thus forming a direct course for the urine to escape, can be relied upon. In instances where the urethra is completely torn across it is usually impossible to introduce a catheter, and here, under all circumstances, the line of action is evident enough. It may be objected that where a catheter can be introduced incision is unnecessary, but it must be remembered that the constant presence of such an instrument in the bladder is no safeguard whatever against the occurrence of extravasation, whilst its continual pressure on the swollen and inflamed urethra at the part injured is likely to be followed by sloughing of the canal and a proportionate extension of any subsequent stricture.

In reference to the third conclusion, that incision diminishes the risk of stricture, or moderates the extent of such a formation, it may be stated generally that the worst forms of stricture are commonly those following laceration of the urethra; and when we consider the circumstances under which such wounds heal where no artificial vent is provided, this is not to be wondered at. The wound in the urethra is of a more or less lacerated character; it heals under the irritating influence of constant contact with urine, and an inordinate amount of plastic exudation is usually thrown out around the wound. On the other hand, in cases where the urethra when lacerated is opened by perinæal incision, we have still the original injury, but the circumstances are more favourable for limiting action. Free vent is here given for all matters of an irritating nature, and the exudation lymph is merely sufficient for coating over the incised tissues.

Longitudinal incisions of the urethra are not ordinarily followed by stricture. We have an illustration of this in the operation of lateral lithotomy, where the formation of a stricture following the incision into the urethra is a rare event, and when it does occur, is probably traceable to some laceration or contusion of the passage in extracting the calculus.

The following case may be mentioned as a further illustration of this observation:—

In 1866, a patient was admitted into the Northern Hospital, under my care, suffering from an incised wound of the perinæum, received in the course of what was described to me as "a free fight," further details of which were not to be had.

The wound was just behind the scrotum, and was very similar in appearance and direction to that which a surgeon would make in opening the urethra in this position. It was a little to one side of the raphé, and was an inch and a half in length. Some hours elapsed before the patient was brought to the hospital. I found the wound plugged with lint, and blood was issuing from the penis. As the bleeding was free, I had to enlarge the wound to enable me to secure what I believe was the transverse perinæal artery. This was sufficient to arrest the hæmorrhage. I could put my finger into the urethra, which had been very neatly incised for about an inch along its floor. Urine escaped freely through the penis, and partly through the wound, for some three or four days. After this time the urine was passed naturally, and the wound closed. I saw the patient about two years afterwards, when he had no sign of stricture, or other indication, beyond the scar, of the injury he had received.

In the after-management of cases of ruptured urethra great attention should be paid to cleanliness, to secure which the wound should be syringed out at least twice daily with weak carbolic lotion. I know of no better application to the perinæum than tenax; it absorbs the discharge and acts as a disinfectant. After the lapse of a few days the introduction of bougies along the whole length of the urethra should be commenced, and continued at regular intervals. When the wound has healed, the patient should be instructed how to introduce an instrument for himself, the use of which he should continue, to prevent any contraction of the urethra taking place.

The following case of rupture of the urethra was attended with a state of priapism, which lasted for three weeks :—

W. M., aged 22, was admitted into the Royal Infirmary, under Mr. Harrison's care, on October 29, 1879, with an injury to the penis and scrotum, caused by a kick four days previously. He stated that on receiving the injury he passed some blood from the urethra, and has continued to do so. On admission there was considerable ecchymosis of the penis and scrotum, and the penis was erect. From an examination of the urethra with a bougie, a partial rupture was made out four inches from the orifice. The patient was kept in bed, and evaporating lotions were applied. A most marked state of constant priapism continued for three weeks, when it gradually subsided, and the patient left the Infirmary on Dec. 1st, 1879. The notes of this case were taken by Mr. A. Stewart.

THIRTEENTH LECTURE.

Perinæal Fistulæ and their Treatment.

Following upon the consideration of urinary abscess and injuries to the urethra, it will be convenient to consider certain results which these sometimes occasion.

I have had, within a comparatively recent period, sufficient varieties of urinary fistula, in persons applying for treatment in my wards, to enable me to occupy your attention for one lecture in the description of some of the forms of these openings and the modifications we find necessary in their treatment.

A urinary fistula may be defined as a sinus communicating with the urethra or other part of the urinary passages through which urine escapes. Like all other sinuses communicating with natural passages, it is often exceedingly difficult to heal, and therefore every care should be taken to prevent its formation, or in moderating the inconvenience it is likely to occasion, should our worst anticipations be realized. These fistulæ are consequent upon either suppurative inflammations in the neighbourhood of the urethra or lacerations of the urethra by means of which the stream of urine is deviated from its normal course; and this includes not only accidental lacerations, but wounds that are necessary in the performance of certain surgical operations in this region. In view, then, of the causes of urinary fistulæ, there are certain rules which experience indicates as being likely to aid in diminishing the chance of their formation, or the difficulties connected with their treatment.

The first rule is, that all suppurations about the urethra should be evacuated at the earliest possible moment. The clean direct cut made by the surgeon is more likely to heal kindly and quickly than the burrowings of a suppuration which has to make its way through structures presenting unequal powers of resistance, such as fasciæ and skin; and the second rule is, that where the urethra has necessarily to be incised, say for a stone which cannot otherwise be got rid of, the incision should be made in the way, and, if possible, at the place, in which the probabilities of its healing without trouble are the greatest; and this, speaking generally, is certainly not in the penile portion of the urethra.*

Urinary fistulæ are met with in all parts of the urinary tract from the kidneys downwards. Such

---

* Bearing upon this point, and no doubt dictated with the view of preventing the formation of a fistula, I find the following remark in an old writer, which I think worth repeating:—" If a small stone be lodged in the urethra near the glans, it may often be pushed out with the fingers or picked away with some instrument; but if it stops in any other part of the channel it may be cut upon without any inconvenience; the best way of doing it is to pull the prepuce over the glans, as far as you can, and then, making an incision the length of the stone through the teguments, it may be turned out with a little hook or the point of a probe; the wound of the skin slipping back afterwards to its proper situation, and from the orifice of the urethra, prevents the issue of the urine, and very often heals in twenty-four hours."—*The Operations of Surgery.* Samuel Sharp, F.R.S. Seventh Edition. 1758.

sinuses, when connected with the kidneys, are usually the result of acute calculous pyelitis. In one instance of this nature, which came under my observation, the patient had had for two years a sinus in the lumbar region, through which urine in small quantities and pieces of calculi occasionally passed.

Fistulous openings into the bladder are far more frequent; these, for the most part, are of a malignant nature, and, when communicating with the lower bowel, are remediable only by colotomy. In the case of a female patient, from whom Mr. Bickersteth removed a large stone by the supra-pubic operation, there was a sinus opening in the buttocks and communicating with the bladder, through which urine was discharged in considerable quantities. Such a complication as this, is however, exceedingly rare in connection with stone in the bladder.

Then we have fistulous communications with the bladder resulting from injury done to the parts during the process of parturition. I merely allude to these; it is to those fistulous openings which show themselves on the perinæum, scrotum, and penis that I purpose referring to in detail.

Let me take an example of the simplest kind. A patient with a stricture has some localized inflammation taking place behind the obstruction; this suppurates, and the matter is discharged either spontaneously or by an incision, we will say, through the perinæum. The discharge of matter is followed by that of urine, which continues to flow through the artificial opening, in greater or lesser quantities, at every act of micturition. So long as the stricture remains untreated, and urine meets with an obstacle to its escape through the natural channel, so long does the inconvenience continue. No plan of treatment specially directed towards the closing of the fistula can possibly be successful until the stricture in front of it is completely dilated. This is the first principle in the treatment of this complication, and to effect the dilatation of the stricture the same rules are applicable as those I have referred to in speaking of the circumstances which indicate the selection of the one proceeding or the other.

In a certain proportion of cases of urinary fistula, all that it is necessary to do is to treat the stricture, to remove all impediment to the escape of urine *per viam naturalem*, and the sinus closes.

I remember a case being admitted into the Infirmary, which shows the importance of dilating the stricture before doing anything else, where an endeavour had been made to close a fistula by the injection of an irritant into it without any regard being paid to the stricture. The patient had been in the habit of passing his urine almost entirely by the sinus. The injection of this irritant led to inflammatory swelling. which blocked up the fistula, and, as might have been expected, brought on complete retention of urine. This was speedily followed by swelling of the perinæum, scrotum, and lower part of the abdomen. Several hours elapsed before his admission into the Infirmary, by which time the scrotum had become almost gangrenous from extravasation. The patient being etherized, I laid open the perinæum, and on a small staff divided a hard stricture, making other incisions in the scrotum and adjacent parts, wherever the presence of extravasated urine rendered it necessary. The patient remained in a very critical state for several days, the tongue being dry and the pulse small and frequent, requiring a liberal allowance of stimulants, milk, and beef-tea. Though almost the whole of the scrotum sloughed away, the patient eventually made a good recovery.

If dilatation of the stricture is not sufficient to bring about closure of the fistula, I recommend an expedient I first saw practised by Dr. Gouley in New York, and which I have on more than one occasion successfully advised—that is, I direct the patient on every occasion he micturates to close up the orifice of the fistula with his finger, and to be careful by the exercise of a sufficient amount of pressure to prevent the escape of a single drop of urine by the false route. Sir Henry Thompson narrates a case where a patient was soundly cured of a vesico-intestinal fistula by causing him to lie on his face when passing water. I have never found any advantage from the retention of a catheter in the bladder with the object of curing a fistula, as the urine invariably makes its way into it by the side of the instrument, but I have seen a fistula completely cured by the patient passing for himself a catheter on every occasion of micturating.

When a fistula has been in existence for some time its walls become thickened, and it is not found amenable to the simple expedients I have referred to.

Under these circumstances, I believe that Brodie's plan is about the best to employ. This consists in stimulating the bottom of the sinus by the occasional introduction of a small piece of nitrate of silver, at the same time retarding the healing of the *orifice* of the sinus (which is more inclined to heal than the bottom of it towards the urethra) by lightly touching it occasionally with potassa fusa. For the same purpose the actual cautery, applied by means of a wire, has been recommended.

In a certain proportion of cases, even where the stricture has been most completely dilated, the fistula fails to heal in spite of the employment of the means I have mentioned. The selection of further expedients will greatly depend upon whether the fistula is attended with a loss of substance or not. And, first, I will refer to those cases in which, though the fistula fails to heal, there is no appreciable loss of substance.

The most important and noteworthy means for effecting closure of a fistula undoubtedly consists in diverting the stream of urine and preventing it under all circumstances entering the false route. The simplest expedient for diverting the stream of urine is that which has been frequently resorted to in the cure of penile fistulæ; we know how difficult these are to close compared with those openings which are made in the urethra for a variety of purposes from the perinæum, which generally speedily heal.

Some years ago, a boy was admitted into the Infirmary, under my care, with a stricture just at the angle where the penis and scrotum joined; it had followed the retention of a metallic catheter in the bladder, which was considered necessary for a rupture of the urethra. Various expedients were tried to close the fistula, but without avail.

There being little or no contraction of the urethra, I introduced a grooved staff and performed the median operation, as if for stone. I then pared the edges of the penile fistula and closed it with wire sutures. The effect of introducing my finger into the bladder through the perinæal incision was so to overdilate the prostatic urethra as to cause the urine, for some days, to be passed incontinently by the wound. During the interval the penile fistula completely healed, and within a fortnight from the operation the perinæal opening also closed, and the patient left the Infirmary with his urethra sound and able to micturate normally.

An extension of this method of treating urinary fistulæ has been adopted with complete success by my colleague, Mr. Banks. I shall refer to the case, not only as illustrating the principle I am at present discussing, but as furnishing us with a practice which is likely to be exceedingly useful in remedying one of the most distressing conditions connected with the urinary organs.

Mr. Banks's case was one of urethral fistula, with an opening in the rectum and another in the perinæum, the intervening tissue being unhealthy and undermined. The treatment adopted was to lay these openings into one, the doing of which disclosed a large aperture in the urethra, through which the urine escaped. All kinds of treatment proving futile, Mr. Banks resorted to the following expedient:—Passing a trocar and cannula into the bladder through the perinæal chasm, he tapped the bladder into the rectum, thus reversing the usual order of proceeding. Then through the cannula he passed an india-rubber winged catheter, leaving the winged end on the floor of the bladder and the other hanging out through the anus; the edges of the deep perinæal cleft were then pared and accurately approximated. After the operation the patient passed his urine through the anus by means of the india-rubber catheter. Within four months of the operation the patient was able to leave the Infirmary and resume his occupation, with his urethra, as it is expressively stated, as "tight as a drum."

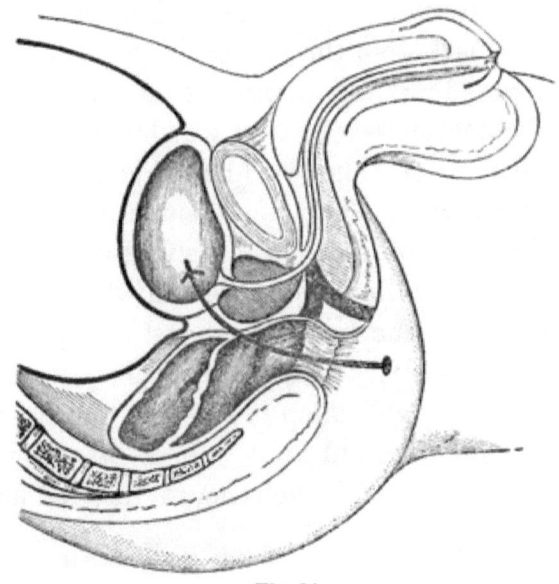

Fig. 24.

I am indebted to Mr. Banks for his permission to reproduce the woodcut illustrating his original paper.

* *Edinburgh Medical Journal*, June, 1878.

The practice of treating these fistulæ by diverting the stream of urine is further illustrated by the employment of perinæal section, which has for its object the division of the stricture and the providing of a single outlet for the urine.

Of the advantages of the proceeding you have seen three instances in my ward during the last year, and these illustrate the performance of the operation under its two different aspects—namely, where an instrument can be passed into the bladder, and where it cannot. The performance of the operation under each of these circumstances has already been fully discussed in my lecture on external urethrotomy. Let me, however, make the remark, that the number of such cases where it was supposed no kind of instrument could be passed into the bladder has been very sensibly diminished by the employment of Gouley's filiform bougies and tunneled instruments. Their successful use in all cases of this kind, excepting where the distal urethral opening has become obliterated by reason of all the urine escaping through the fistula, is almost certain.

And now I will take one or two illustrations from my wards of the application of external urethrotomy to perinæal fistula. And first, where an instrument could be passed through the stricture into the bladder.

J. B., aged 41, was admitted into the Infirmary, under my care, on August 1st, 1878.

His history was that of a stricture following gonorrhœa, a perinæal abscess, or rather abscesses, and urinary fistulæ, through which the greater proportion of his urine had for some months been passing.

On his admission, I found a tight stricture in the sub-pubic urethra, through which a No. 3 English bougie could just be passed; there were three fistulæ, one perinæal and two scrotal.

The patient was very intolerant of the passage of a bougie, and I determined to perform external urethrotomy as soon as I could get my No. 6 grooved staff fairly into his bladder. This took me three weeks to do, when I was enabled to accomplish my object, and completely divide a long stricture from the perinæal wound. This permitted me to pass a full-sized instrument throughout his urethra, so satisfying myself that I had removed all resistance.

As I had the three fistulæ now to deal with, with the view of completely draining them, I passed a gum-elastic catheter from the perinæal wound into the bladder, through which all the urine escaped. This I retained for forty-eight hours, by which time I had diverted the whole of the urine from the fistulous tracks, and through the perinæal opening. Not to weary you with details of treatment, in five days after the operation I commenced to pass bougies of a full size, which I continued to do until the perinæal wound had completely closed. In six weeks, and without any special attention, the fistulæ and the perinæal wound had thoroughly healed, and the patient left the hospital with instructions to pass for himself a full-sized bougie, which he had speedily learned to do.

As an instance of the successful treatment of a stricture which I, as well as others, found to be impassable, I believe from obliteration of the distal aperture, and complicated with no less than nine fistulæ, I cannot do better than record the case of the patient at present in hospital, and who is just about to leave:—

Thomas D., a sailor, aged 39, was admitted into the Royal Infirmary, under my care, on May 30th, 1879.

His history was that of a gonorrhœa nine years ago,

subsequent stricture and urinary abscesses, in addition to two injuries to the perinæum resulting from falls.

On admission into the Infirmary, the scrotum and perinæum were swollen and brawny, and riddled with no less than nine fistulæ, running in various directions, including one which distressed him more than the others, as it opened on the surface of the abdomen amongst the pubic hairs. The state of the parts well represented, during micturition, the rose of the typical watering-pot.

On examining him on several occasions, I could never succeed in passing the finest instrument into the bladder. I question whether the urethra was at all permeable, for, in addition to the above-mentioned circumstance, I was unable to ascertain that any urine escaped during micturition otherwise than by the fistulous tracts.

In considering what to do for the patient, I felt that his state was so deplorable that almost any risks were justifiable which offered a reasonable prospect of improvement. There was no evidence that his kidneys were seriously unsound, and, even had they been so, I should have been disposed to believe that the sense of relief an operation might afford would more than compensate for the shock occasioned by it—just as we see the most miserably hectic patients spring, as it were, into existence when the irritation of a diseased joint is removed. I therefore determined to perform perinæal section, and to give the patient one direct channel by which he could discharge his urine. The only difficulty, so far as the operation was concerned, was that I had to do it without the guide of an instrument passed into the bladder, and upon a urethra necessarily contracted behind the stricture, by reason of the fistulous openings. However, all this was safely accomplished by keeping my incision in the median line and with the apex of the prostate as an unerring guide. The result was all that I could have wished; the relief afforded by the one opening was most marked, and the patient made very satisfactory progress. At the time of the operation I believe I opened up the occluded distal orifice of the urethra. Presently I got in a fine gum-elastic instrument, and by means of continuous dilatation the calibre of the strictured urethra was brought up to a No. 9, English scale. All the fistulæ closed. Much of the hardening disappeared, and the patient left the Infirmary able to micturate naturally—a striking contrast to his state on admission. Such, then, is an illustration of the results we may obtain in impassable strictures, complicated with numerous fistulæ, by external urethrotomy. I am indebted to my dresser, Mr. R. P. Sykes, for the notes of this case.

As somewhat differing from the case just recorded, but bearing upon it, I would cite the particulars of the following, in which there was occlusion of the urethra consequent on a wound:—

This patient, a boy aged 11, I saw with Dr. W. Little, formerly of Everton. He had been crushed by a carriage, and had evidently sustained some severe injury to the pelvis, as well as to the urethra. He had complete retention, and we were unable to pass the catheter further than the deep fascia, where the urethra appeared to be wholly severed. Under these circumstances we agreed that a free perinæal incision was required, and this I accordingly made. A considerable quantity of urine and extravasated blood escaped. A fracture of the pubic arch was also discovered. For several weeks the patient remained in a very precarious state, as the injury was followed by an acute attack of peritonitis, and for weeks all urine escaped by the wound. As soon as the patient's health permitted, I attempted to establish the continuity of the urethra; the canal having been completely severed, the distal end had closed. This was a very troublesome affair, but eventually it was accomplished, and the perinæal wound healed. I had the patient under observation for nearly eighteen months, and when I last examined him, though the urethra admitted a full-sized bougie, yet at the site of injury the tissues were rough and cicatricial, so that I fear there may be some permanent contraction.

I have, lastly, to notice those forms of perinæal fistula attended with more or less destruction of the tissues adjacent to the urethra, the simple closing in of which would obstruct the urethra. To remedy these, a plastic operation is necessary, in addition to the application of those principles to which I have referred, and which are equally essential to successful treatment. I might occupy your time at considerable length by a description of the various modes of transplanting tissue for this purpose which have been adopted; the most important of these you will find detailed in the text-books, which will prove trustworthy guides when combined with that resource in adaptation which springs from a sound application of surgical principles. Some of the most original designs in the plastic surgery of the perinæum and penis will be found in almost any of the modern French treatises relating to this subject.

FOURTEENTH LECTURE.

Foreign Bodies in the Urethra and Bladder—Action of Urethra—Illustrative Cases—Use of the Lithotrite as an Extractor—Foreign Bodies in the Female Bladder.

Through accident or by design foreign bodies occasionally become lodged within the urethra or the bladder.

Amongst the miscellaneous articles that have been found in one or other of these positions I can recall to mind pins, needles, wires, a lucifer match, a knitting needle, a slate pencil, a feather, a bulb-headed grass, pieces of catheters and bougies, a whole bougie, and a pencil-case; but, taking the experience of others, this repertoire might, I expect, be considerably extended. Most of these articles have been introduced for the purpose of acting upon the penile portion of the urethra—for reasons often best known to the patients themselves,—and, having slipped from their grasp, have made their way into the bladder. You will find, on reference,[*] a correspondence between Sir Henry Thompson and Mr. Christopher Heath bearing upon this subject, the latter stating, "I have noticed in perfectly healthy urethras that there is a constant

[*] *The Lancet*, vol. ii., 1866.

vermicular contraction of the wall of the canal, apparently passing towards the bladder; and this accounts for the well-known fact that foreign bodies in the urethra tend to pass in that direction." Sir Henry Thompson, on the other hand, maintained that the vermicular movement " is precisely in the opposite direction; and also that foreign bodies have a strong tendency to pass outwards to the meatus, and not inward to the bladder."

My own impression is that the vermicular action of the urethra is an ejaculatory one, and that a foreign body is only forced towards the bladder when, in itself, it presents some obstacle to its passage outwards. A piece of bougie placed within the urethra, with its anterior extremity broken and uneven and its posterior end smooth, is sure by vermicular action to be forced in a direction *towards* the bladder by the very efforts that are made by the urethra to expel it. Just as when the movements of the intestines are interfered with by a portion of the gut becoming strangulated, the contents of the canal are, by vermicular action, thrown backwards, instead of being propelled in a direction onwards to the natural outlet.

I will proceed to narrate some cases where foreign bodies have been passed into the urethra, as serving to illustrate certain points in practice which they suggest.

In 1861, Mr. Stubbs saw a youth, aged 16, who was suffering from some induration in the ischio-rectal fossa and perinæum. On inquiring into his history, he found that eighteen months previously the patient had passed into his urethra a good-sized needle, which, having slipped from his grasp, disappeared. Not liking to mention it to anyone, he had refrained from seeking surgical advice. He appears to have suffered very little inconvenience from his accident up to within a week of his being seen, when he had some pain about the perinæum and difficulty in micturition. As fluctuation could be felt in the ischio-rectal fossa, an incision was made, and some matter evacuated, but no needle could be felt. Mr. Stubbs introduced his finger into the rectum, where he could distinctly feel the sharp point above. He then incised the perinæum in the median line, and removed the needle represented in Fig. 25.

Fig. 25.

The patient recovered rapidly, and the opening entirely closed, no urine escaping through it. The needle, apparently, was more than half covered with a deposit of lithic acid. (This specimen is amongst the calculus collection in the Museum of the Liverpool Royal Infirmary School of Medicine.)

The needle appears to have been introduced blunt end foremost, and so to have made its way into the bladder, where, in the course of time, it became largely coated with calculous deposit. Escaping from the bladder, it became lodged in the perinæum, and was extracted in the manner described.

It is interesting to observe the calculous incrustation that took place on it, as explanatory of the formation of stone,* this being one of the ways in which

* In the Museum of the Royal College of Surgeons is a calculus (A 126) from the human bladder, having a slender piece of steel for a nucleus. In reference to the specimen the catalogue contains the following note:—" The deposit of uric acid, or any other substance except the earthy phosphates, upon foreign bodies in the bladder, is exceedingly rare."

a calculus is formed. We have had numerous instances of this. In one case where I performed lithotomy, the nucleus proved to be a piece of bone. Mr. Wilkes, of the Salisbury Infirmary, records a case of the same nature in Part 45 of the *Proceedings of the Royal Medical and Chirurgical Society of London*, which presents special points of interest.

In 1865, Mr. Stubbs removed, at the Royal Infirmary, a urinary calculus from the bladder, the nucleus of which was a piece of bougie about one inch in length; in this case, curiously enough, there were no symptoms whatever of stone in the bladder. The patient was suffering from stricture of the urethra, for the relief of which perinæal section was decided upon. Mr. Stubbs performed the operation in the usual way, and on an instrument being passed into the bladder a stone was felt, and without much difficulty removed through the perinæal wound. The patient had been in the habit of introducing a bougie, and remembered, three months previously, a piece breaking off; but as no untoward symptom resulted, he thought nothing further of it.

Fig. 26 represents a phosphatic calculus, from the Museum of the School of Medicine, the nucleus of which is a piece of metal.

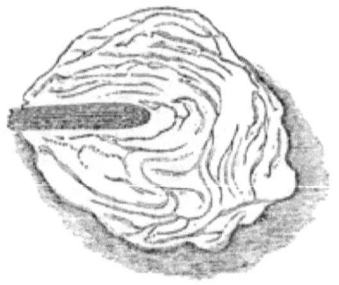

Fig. 26.

Amongst the more remarkable objects that have found their way into the bladder is a bulbous grass.

This case came into the Royal Infirmary in 1865, under the care of Mr. Long. The patient supposed that he was suffering from stricture, to remedy which he was in the habit of passing different materials into his urethra. On this occasion he selected the ear and stalk of one of the grasses, which was introduced readily enough, but could not be withdrawn. Further efforts on his part only made matters worse, the ear being forced from the urethra into the bladder. When admitted into the Infirmary, shortly after the accident, he was suffering from most acute cystitis. A lithotrite removed a portion of the grass-head, slightly encrusted with phosphates. The symptoms, however, were not abated, and death resulted. On opening the abdomen there was found general peritonitis, the viscera being either adherent or coated with a layer of lymph. Within the abdominal cavity were four or five pints of turbid serum, which exhaled an ammoniacal odour. Upon examining the bladder, a large head of one of the grasses, covered with phosphatic deposit, was found impacted within it.* The stalk, which was stiff and resisting, had made its way through the fundus of the bladder, and protruded into the peritoneal cavity. The pelvic cellular tissue was infiltrated with purulent matter, having a urinous odour. The inflammation extended up the ureters to the kidneys.

The history of the case was not obtained without

---

* What I have described as a "grass-head," Dr. W. Carter, the Lecturer on Materia Medica at the Liverpool School of Medicine, kindly informs me is the spikelet of the meadow fox-tail grass (*Alopecurus pratensis*).

considerable difficulty, and what had actually been inserted into the urethra was almost a matter for speculation, which the introduction of the lithotrite only incompletely determined.

The sketch of this exceedingly interesting case (Plate A) is from the specimen in the School of Medicine Museum.

A somewhat similar case is recorded, in which Mr. Heath, of Manchester, removed by lithotomy about three inches of the stem of a sage-plant, with a thick coating of triple phosphate at the distal end.*

The next specimen I will show you is a needle, armed with a knob of sealing-wax, which was passed into the urethra, for some imaginary complaint, by a young gentleman, and ultimately made its way into the bladder. After remaining there for some twenty-four hours, it appears to have been forced out into the perinæum, whence it was removed through an incision, by Mr. Swinden, of Wavertree, who kindly presented me with it. The sketch (Fig. 27) represents

Fig. 27.

its actual size, from which it will be seen that it measures nearly three and a half inches in length.

In 1864, a case was related at the Liverpool Medical Institution by Mr. Hamilton, in which, at the Southern Hospital, he had removed from a man's bladder portions of calculous concretion formed on a

* *Manchester Medical and Surgical Reports*, vol. ii.

PLATE A.

feather, which had been passed by the patient for the relief of a stricture. Here lithotomy was performed, as, from the nature of the stricture, lithotrity was impossible.

I will now pass on to notice a case in which I removed a foreign body from the bladder by means of the lithotrite, and which testifies to the value of this instrument under such circumstances. I am indebted to my dresser, Mr. I. Holmes, for the notes.

W. O., aged 38, a militiaman, was admitted into the Liverpool Royal Infirmary on May 22nd, 1877. His statement was to the effect that, on the previous night, when under the influence of liquor, a pencil-case had been introduced up his urethra by a prostitute, in whose company, together with others, he had been. He did not appear, however, to have discovered anything amiss till the following morning, when certain uncomfortable sensations in the region of his bladder made him come to the conclusion that the lost pencil must be there. From his manner I was first inclined to think that the man was insane, but on hearing that the surgeon of his regiment had discovered the existence of a foreign body in his bladder, and had sent him to the Infirmary, I at once proceeded to examine him.

Upon examination, the foreign body appeared to be lying obliquely, partly in the bladder and partly within the prostatic portion of the urethra. I first attempted to remove it by means of the extractor, which is described in Reliquet's *Traité des Opérations des Voies Urinaires*, and known as the instrument of Messrs. Robert and Collin; but failing, a lithotrite was passed. By this the pencil was carried on completely within the bladder, where it was seized transversely. In this position it was impossible to extract it; however, by gradually rotating the lithotrite toward one side, whilst the pencil was kept within the blades of the instrument, I succeeded in reaching one end, when the pencil was removed, point foremost, without any further difficulty or damage to the urethra. The exact size of

Fig. 28.

the pencil-case is represented in Fig. 28. The patient was placed in bed, and a linseed poultice applied over the abdomen. During the afternoon and night he passed urine naturally, and on the following day appeared in no respect the worse for what had been done. He was kept in the Infirmary until May 24th, when he went out on leave, but did not return.

Referring to the manipulation employed in extracting the pencil, I should say that similar means had been successful in another case which had come under my notice only a short time previously, in which I had succeeded in removing from the bladder a piece of gum-elastic bougie that had been accidentally broken in the urethra. In the instance I have just recorded the extractor failed, because at first the pencil was firmly impacted, and therefore could not be made to rotate. Had I not succeeded with the lithotrite, I should have had recourse again to the extractor, which would then probably have been successful, inasmuch as, the pencil being fairly within the bladder, rotation would have been practicable. The difficulty in removing foreign bodies from the bladder, such as pieces of bougie, is due to their being generally seized by forceps, or the lithotrite transversely. This difficulty Messrs. Robert and Collin have endeavoured to overcome by the use of an instrument something like a lithotrite, the blades of which are so arranged that on seizing a body, such as a piece of bougie, it is rotated, and its long axis made to correspond with the course of the urethra. The instrument is shown here from Reliquet's work, to which allusion has already been made. (Fig. 29.)

The second case of this kind to which I shall refer is one in which, by means of the lithotrite, I removed from the bladder of a middle-aged man a No. 3 bougie, twelve inches in length. The bougie had been introduced by a surgeon as a conductrice to a urethrotome, with which it was intended to divide a stricture by internal section. Unfortunately the bougie separated from the urethrotome just beyond the point where it was attached by means of a screw. The surgeon at once ruptured the stricture by Holt's method, and left the bougie in the bladder for extraction on a future occasion. I saw the case fourteen days after the accident. As the urethra would by this time admit a No. 12 bougie, I had no difficulty in introducing the lithotrite and extracting the bougie. This I seized about the centre, and brought out doubled; it being soft and of small size, the removal was accomplished by merely gentle traction. The patient

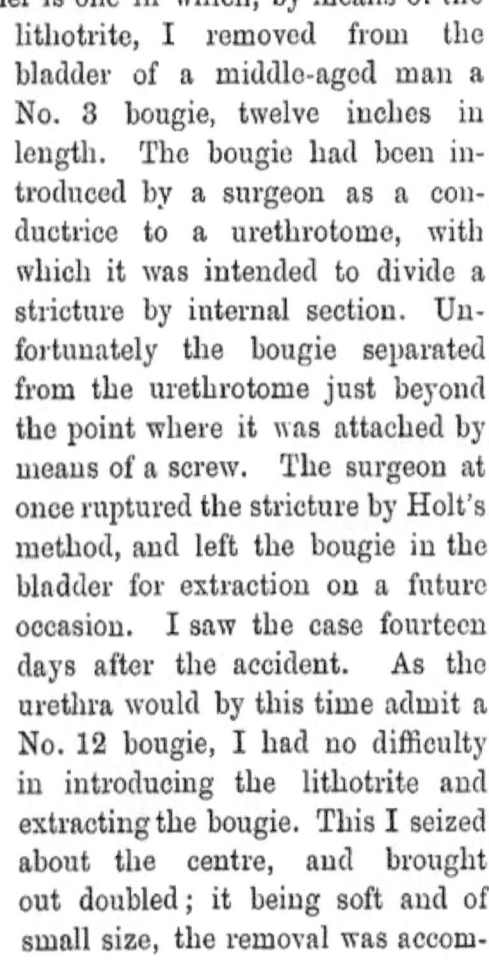

Fig. 29.

recovered without a bad symptom. The bougie appears to have remained curled up in the bladder; no calculous deposit was observed upon it, although it had been retained for a fortnight.

This case points to the necessity of care being exercised in properly securing the connecting links between the urethrotome and the guide. Additional means of security have, I understand, been taken in the construction of the kind of instrument used on this occasion, to obviate the occurrence of such an accident as I have described, which might have given rise to much more serious consequences.

Mr. Lund, of Manchester, records a very similar case, where he removed a bougie from the bladder by the lithotrite. In this instance, also, it was complicated with stricture, for which Holt's operation was performed previously.*

An extremely interesting case, where a piece of bougie remained in the bladder for five years, and was successfully removed, encrusted with phosphates, by lithotomy, is recorded by Mr. J. W. Baker, of Derby.†

The female bladder is also occasionally found to contain various foreign bodies.

Some years ago, by means of a pair of narrow dressing forceps, I removed a bodkin, which had been introduced by the patient for the purpose, it was alleged, of extracting a piece of gravel from the urethra.

A remarkable case of this kind was narrated by

---

* *Liverpool and Manchester Medical and Surgical Reports*, 1873
† *British Medical Journal*, Dec. 5, 1874.

Dr. Grimsdale, at the Liverpool Medical Institution, in 1865, where he had removed a calculous concretion, formed on a large hair-pin, from the bladder of a young lady, aged fifteen years. Removal of the foreign body was effected with forceps, after rapidly dilating the urethra with a Weiss's dilator. On the second day after the operation she was able to pass water voluntarily; recovery followed, the patient possessing full power over the bladder. In this instance there was some tumefaction above and to the left side of the symphysis pubis, as if an abscess were impending. It is probable that the foreign body might have been expelled in this way had not its removal been effected by surgical interference.

The Museum of the School of Medicine also contains a hair-pin (E 22) which was removed by Mr. Bickersteth, by rapid dilatation, from the bladder of a female. Hair-pins appear to be rather favourite articles for passing into the female bladder, as I see another case of this kind is recorded by Dr. Johnstone.*

As, in cases of stone, operations on the female bladder where the urethra is incised are apt to be followed by incontinence of urine, rapid dilatation of the urethra should be employed for the removal of all small bodies. Where the calculus, or foreign body, is too large for extraction entire in this manner, lithotrity may be advantageously combined.

* *British Medical Journal*, Sept. 27, 1879.

FIFTEENTH LECTURE.

IRRITABLE BLADDER.

The term "irritable bladder" has been used to express a disease rather than a symptom, and hence some confusion has arisen in the application of therapeutics to the condition upon which this perverted function depends.

It is as important to define accurately the circumstances which give rise to this irritability in the bladder as it would be if we were to proceed to discuss, with a view to their treatment, the causes producing a similar symptom in any other organ, such, for instance, as the eye.

To prescribe correctly for an irritable bladder, the same careful consideration of the pathology of the subject is required as is needed in the more obvious illustration I have taken.

For the sake of clearness let me define what we understand by the term, and then I will endeavour to notice under what circumstances we meet with the condition. In this way I shall hope to apply, to some purpose, principles of treatment which may be of service in enabling us to remedy not only a very distressing, but a very common symptom.

Let us clearly understand, then, that in cases

of irritability of the bladder we do not commit ourselves to attempting to correct this ailment until we have asked the question—What is the cause of it? It will be quite time enough to be empirical in our advice when we have failed to satisfy ourselves upon this point.

The term "irritability of the bladder" we understand to mean that the act of micturition is performed unnaturally often.

I do not attempt to qualify this somewhat broad definition by any statement as to the number of times a healthy person should micturate in the twenty-four hours. There are variations dependent upon circumstances and individual peculiarities which would render such an attempt almost ridiculous, and therefore unless a person were in some manner inconvenienced by the frequency with which the act was performed, I should not hold that he was suffering from irritability. Irritability of the bladder is traceable to one or other, and sometimes to more than one, of the following conditions: nerve, habit, reflected action, structural diseases, including tumours and calculous disorders, and abnormal condition of the urine.

In a given case of irritable bladder, if you will go through these headings carefully and systematically, you will probably succeed in doing that which others may have failed in elucidating. If you do not take up the enquiry in some such methodical manner, it is not at all improbable that you will miss your mark, and then have—as is quite within your right, if your system, whatever it may be, fails—to fall back upon empiricism.

Now, you may possibly ask, What do you mean by "nerve" as determining an irritable condition of the bladder? I mean just what you do when you make use of the term in its ordinary acceptation — "a nervous man." You will find a certain proportion of cases of irritable bladder simply due to this condition —not to disease, but to the nerve tone of the individual. I have known many persons, in anticipation, for instance, of a long railway journey, go on for days previously micturating every few minutes, in view of an imaginary inconvenience to which they might be temporarily exposed. These individuals are simply nervous upon this point, and acquire a habit which sometimes becomes permanently established. "Nerve" and then "habit" will produce a state of urinary irritability which it is exceedingly difficult to throw off. For many years I was constantly consulted by an elderly gentleman who suffered in this way; the dread of the fortnightly journey to London for business purposes manifested itself thus until a habit was acquired which seriously threatened the patient's health, and yet there was nothing but these combined influences to explain matters. All sorts of expedients only seemed to add to his distress for fear of their failing, until on the introduction of the Pullman system of railway carriages I suggested their use, which was at once followed by a total cessation of this unpleasant symptom.

You have often no other objective symptoms in these cases to guide you; the history of the patient, the circumstances influencing him, together with that important assistance which is afforded by the method of investigation known as the process of exclusion, will be your guide. Before coming to the conclusion that the case is one of "nerve," or of "habit," or of both, you must fortify yourself with the assurance that there is a complete absence of all signs of structural disease. Having satisfied yourself as to the cause, what is to be done for these patients? as, from the varying circumstances attending these cases, you cannot cure them all by a Pullman carriage. You will employ the same principles of treatment as you would, with obvious modifications, in any other like form of nervous disorder. In the first place, you will explain to the patient his position, and give him substantial reasons for the view you are taking. The employment of a little unvarnished common sense in these cases, shaped in reference to the peculiar dread the patient may have, and which has induced the irritability or added to it, will often avail much. As soon as your patient finds out that you know as much about his unpleasant sensations as if you had the complaint yourself, you will be able to exercise an influence for good over him which he will not be long in acknowledging. In no other way will you be able to give your patient the assurance he requires.

The strange vagaries one meets with sometimes in connection with disorders of the urinary organs may possibly occasionally try your patience. It was only a short time ago that a patient presented me with a very elaborate table, in which he had positively noted by his watch the number of seconds he took to perform each act of micturition; these he worked out mathematically, and furnished me with the ratios of time to quantity of urine, and asked me to explain certain differences which appeared to him very terrible and ominous. Yet he was sane, and only asked for information which I took pains to give, and thus to satisfy him. And so you must do in all these allied affections, for affections they most certainly are.

In addition, however, these cases require, and are benefited by, all those medicines whose power in toning the nervous system is so well known. I speak generally of the preparations of iron, nux vomica, strychnia, and phosphorus. I have sometimes found the bromide of ammonium, in twenty to thirty grain doses in water, three times a day, act almost as a specific in this condition.

Of the solution of the dialysed iron, which has recently been introduced, I can speak favourably, as strengthening nerve power without disturbing the digestive functions.

On the use of aperients, of baths, and of suitable diet, both in eating and drinking, I need not insist, as these are essential in promoting both nerve and physique generally.

Persons who have had any reason, however slight, for believing that they suffer from stricture, not unfrequently develop irritable bladders. I have seen this follow upon all kinds of misapplied constructions relating to normal acts of micturition; upon the unskilful introduction of instruments undertaken with the object of removing such doubts; or a groundless dread of inability to void urine, inducing a frequency in the act which has eventually resulted in the setting up of irritability of the bladder. It is astonishing how many persons may be completely cured of this symptom by demonstrating to them the ease with which a bougie may be made to enter the bladder.

Irritability of the bladder, dependent upon reflected action, is most commonly met with in children and young persons. Illustrations of this class are furnished by the irritability that occasionally attends the presence of intestinal worms, and similarly I have known the cutting of a tooth in a child produce the same effects. In youths particularly, and even in those of a more advanced age, I have known a constant desire to micturate kept up by an elongated and contracted prepuce. The retention and decomposition of the preputial secretion, by setting up inflammation, indicates this as the probable cause. For the latter condition, the operation of circumcision at once removes the cause of complaint. How it is that irritability of the bladder is caused by the reflected action which a long or tight prepuce sets up, and conversely, why pain is felt at the end of the penis when there is a stone in the bladder, are well illustrated by Mr. Owen in the accompanying Plate.* (Fig. 30.)

Irritability dependent upon structural changes in the urinary organs, growths, and calculous affections, is with these a most frequent concomitant; nor are these causes entirely confined to those diseases of the urinary organs which custom has brought more espe-

* Harveian Lecture. *British Medical Journal*, Feb. 28, 1880.

cially into the hands of the surgeon. Certain forms and stages of purely renal affections do not seldom give

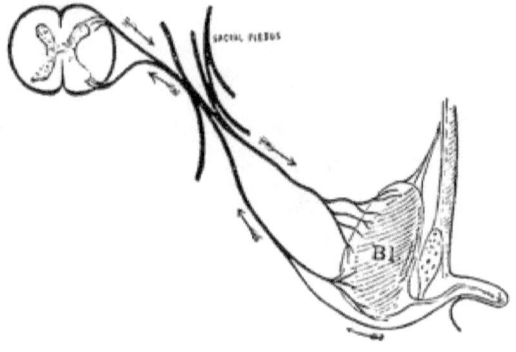

Fig. 30.

rise to this symptom. Whether in these instances the irritation is merely reflected or is directly due to some alteration in the urine, is an inquiry I do not intend entering upon; it is sufficient here to remember that such may be the cause, whatever may be the explanation.

Passing to the bladder, we find that in some period of their course all growths give rise to irritability, in addition to the other symptoms indicative of their presence. Similarly, irritation is provoked by enlargement of the prostate, particularly at the commencement of the disease. At this stage it would be more correct to speak of this as senile engorgement of the prostatic veins, a condition which often precedes and is mistaken for the structural enlargement of the gland with which we are familiar. The irritability connected with prostatic engorgement shows itself chiefly at night. The patient is perfectly well during the day, but as soon as he gets into bed he experiences a desire to pass water, which further disturbs his rest by provoking another call after intervals of varying extent; or it may show itself by inducing a state of more or less priapism, which equally interferes with sleep. Of this I have met with several examples. During the day, as I have already said, the patient is free from irritability; it only occurs at night. In this condition, physical examination with the finger in the rectum, or a catheter in the bladder, frequently fail to detect any signs of prostatic hypertrophy; possibly all that may be noted is a distended or varicose condition of the veins immediately in front of the finger. Provided, as is most usually the case, there is nothing in the state of the urine to account for this, I have found some very simple expedients of service in remedying, if not entirely putting a stop to, this symptom. The wearing of warm socks at night, or the use of a hot bottle to the feet, by determining the blood to the legs, I have often found of considerable service—a fact which leads me to believe that the appearance of this symptom only at night is due to some alteration in the venous condition of the part by reason of the change in position.

Senile engorgement and hypertrophy of the gland not only follow in succession, but the two conditions frequently co-exist. Irritability of the bladder, due to an enlarging prostate, is usually determined without difficulty by physical examination. It will be noticed that the irritability of hypertrophy varies somewhat in the precise mode of its causation. In the earlier form of the enlargement, especially in gouty subjects passing highly acid urine, the irritation comes on immediately after the bladder has been emptied, and the desire will remain for an hour or so until some urine has collected, and a water-bed is as it were interposed between the muscular pressure of the bladder and the tender or gouty prostate. Then there is an interval of repose until urine is again passed, when the same process, accompanied by similar sensations, is experienced. The irritability of the subsequent stages of prostatic enlargement is somewhat different in its character, being due to residual urine plus the chronic cystitis which the pathological state of uncleanliness has engendered. The difference has this import—the former is aggravated by catheterism, as usually practised, whilst the latter is remedied by it, combined with irrigation of the bladder; and conversely, sedatives and emollients give relief to the first-mentioned form of irritability, whilst in the latter, they are, alone, worse than useless—they are disappointing.

Irritability of the bladder in children and adults is a usual symptom of stone, though it varies much both in kind and degree. There is this anomaly in the bladder irritability of stone which has often struck me: in most other forms of irritability a patient gives way to it, at all events, with the prospect of a temporary relief; whilst in stone, on the contrary, it is with the certainty of having his suffering added to until urine collects sufficiently to take the pressure of the calculus from off the mucous membrane.

Some time ago I was seeing a gentleman for irritability of bladder, who, I strongly suspected, was suffering from stone. I had searched his bladder with a sound, and also with a catheter, for a cause, but, like others, in vain. One day I was examining him with a prostatic catheter after I had previously, as I thought, emptied the bladder of all the urine it contained, when, in moving the instrument about, I felt it suddenly pass over something with a jerk, and then, on gently pressing it, it went a couple of inches further in. This was followed by the discharge of about two ounces of urine, which struck me as being unusually milky-looking. There was no bleeding. My friend went home, pondering over a suggestion that I made to him—that his irritability was due to a sacculation, in which urine lodged, and which we had accidentally discovered; at all events, he was immediately relieved, and remained so for forty-eight hours, when the feeling of irritability returned. Having had some experience in catheterising himself, he at once passed the instrument which I had lent him, and having drawn off some water that he had purposely retained, he began cautiously to grope about his bladder, with precisely the same result as had happened to me. The patient came to the same conclusion that I had—namely, that he had a sacculated bladder, and being an ingenious man, he devised a stylet by which he could readily pass his catheter into the secondary receptacle. When I saw him last, accidentally, he informed me that by this way, passing an instrument for himself from time to time, he was completely cured of this irritability, and he believed that the sacculation had almost, if not entirely, disappeared.

Such, then, is an illustration of irritability dependent upon urine retained in a secondary bladder or receptacle. I am afraid you will not find all such cases so satisfactorily remedied. Apart from its extremely interesting nature, the case is worth remembering as exemplifying the kind of irritability you may expect to have under these circumstances.

The diagnosis of sacculated bladder is not always easily made. When suspected, it is not a bad plan, after catheterising the patient, say when recumbent and having emptied his bladder, to alter his position by making him stand, and then seeing if more urine escapes on moving the catheter gently about; or the order in position may be reversed. Guthrie mentions the case of a gentleman in whom the existence of one or more pouches in the bladder was determined by injecting the bladder with warm water; on withdrawing it only a portion could be obtained, and rarely the whole of it, even by any change of position.*

Stricture of the urethra is not uncommonly attended with some degree of irritability. In the earlier form of stricture it is due to disturbed muscular action, but in the latter forms of this disease it is due to those changes in the walls of the bladder to which reference has been made. The bladder becomes so structurally altered as to be unable to perform its function. It is interesting to notice how long the irritable condition of the bladder will remain after the stricture has been

* *Op. cit.*, p. 30.

remedied and the urethra dilated to its natural size; that is to say, how long the small thickened bladder will take to again adapt itself to the ordinary emergencies of micturition. In the majority of individuals it does so adapt itself, but in others it continues to be irritable long after the stricture has been completely relieved. In some cases I have endeavoured to remove this trouble by the use of sedatives, and so to make the bladder more tolerant of its contents. In two or three instances I have tried the effects of a solution of borax introduced into the bladder, so as to exercise a hydrostatic pressure on its walls, by means of the funnel and elastic tubing devised for the purpose of irrigation, and I have found benefit from this.

Lastly, I have had to resort to cystotomy, the irritability remaining, after the stricture has been to all intents and purposes cured, being intolerable to the patient. Twice I have performed cystotomy under these circumstances, following the directions laid down for Allarton's median operation, and thus permitting the urine to escape continuously by the perinæal wound.

It may be asked, What is the rationale of this? How are you going to make a bladder tolerant of urine by not allowing any urine to collect in it for an interval of two or three weeks, or even more? My answer to this is, that the continuance of the irritability after the stricture has been cured is due to the considerable concentric hypertrophy which the bladder has undergone in its constant endeavours to force the urine through the obstruction in front of it. What is the effect of preventing for a time a muscle exercising its power of contractility? Atrophy, or loss of muscularity. This is only an exemplification of an established physiological law; and so it is with the hypertrophied bladder, as I have in practice proved to be the case. In a recent instance of this kind the particulars were as follows:—

W. J., aged 49, was re-admitted into the Infirmary under my care on Feb. 21, 1879, having previously been treated by me for an old-standing stricture with irritability of the bladder. After his stricture had been fully dilated, I told him that it was probable in the course of a short time his bladder would adapt itself to the improved state of the urethra, and again become tolerant. This, I explained to him, was the rule in such cases, and I advised him to maintain his improvement by the occasional introduction of a bougie. In spite of this, the irritability continued, and he was re-admitted to the Infirmary on the date mentioned.

On examination, I found his urethra capable of receiving a large bougie, but his bladder was thickened, and so contracted that I did not think it would hold an ounce of fluid. I tried several expedients, mechanical and medicinal, to remedy this state of things, but without avail. His condition of irritability, both by night and day, was quite as bad as anything of the kind I had ever seen. I therefore resolved temporarily to paralyse the bladder by the performance of median cystotomy, which I did, with Mr. Rushton Parker's assistance, on April 2, 1879. On passing my finger into the bladder, nothing further than what has been noticed could be felt. I should have mentioned that the irritation felt by the patient at the end of his penis was so severe as to have led him, in endeavouring to relieve himself, to scratch his glans until a deep sore was formed, as large as a sixpence, by the side of the meatus. Immediately after the operation the sore began to heal, and on April 15th it is noted by my dresser (Mr. Renner) that "the sore on the glans penis has completely healed." The operation was followed by complete relief, the patient, for the first time for many months, and without anodynes, obtaining sound sleep; in fact, he appeared to be making up lost time, as, in addition to his nights, he spent most of his days in sleep. On April 25th, urine, for the first time, began to pass along the urethra, and by the 30th the perinæal wound had completely healed. The patient left the Infirmary very greatly improved; though passing his urine more frequently than usual, he was quite free from those spasmodic pains connected with the irritability which distressed him so much. I have since heard that the improvement continues. On referring to this patient's temperature chart, I notice that the rise of temperature after the operation was hardly appreciable—additional evidence, I take it, of the immense relief that was thus afforded.

And now I will proceed to notice the irritability that is due to altered and abnormal states of the urine. This is not infrequent, and will necessitate, where there are grounds for suspicion or symptoms to be cleared up, such an examination of this excretion as I have referred to in another lecture. The urine least irritating to the urinary passages is that which most nearly approaches what we have taken as a healthy standard. The low specific gravity of the urine that is passed so frequently and abundantly by hysterical females no doubt causes the irritability of the bladder from which they, under these circumstances, almost invariably suffer. Water is more irritating to those passages of the body over which it is not intended to flow, than a saline solution of some density.

The abundance of uric acid in the urine of the gouty undoubtedly explains the extreme irritability of the bladder and the intense irritation and feeling of weight these persons experience, and refer to the region of the prostate. My opinion is that gouty manifestations in the parts behind the triangular ligament are quite as frequent as the more familiar indication of this diathesis which we meet with in an acute form about the ball of the toe. The benefit that attends the administration of gout remedies in these cases, together with diluents, is most marked. I believe that something similar is seen in persons who frequently suffer from gouty appearances in the skin, such as eczema. I have a patient, undoubtedly gouty, who successively suffers from eczema and irritability of the bladder. Neither is present to any degree at the same time, and the one seems to be the alternative of the other. It is not my intention to follow this inquiry further; I merely wished to remind you how it was that abnormal conditions of urine might explain the symptom we are now considering, and that the correction of this flaw would most probably remove it.

There is a form of irritability—for so it certainly is, though manifesting itself by actions rather than by sensations which are morbid—where the bladder is not under proper control. I allude to the nocturnal incontinence of young children, which may be provoked by any of the causes previously mentioned, and which must be carefully sought for. This state is not to be mistaken for the dribbling or running over of a distended bladder, which, by atrophy of its muscular coat or other similar cause, is prevented expelling its contents. Whenever we hear of incontinence in an adult we must be alive to this, and not allow the actual condition of the bladder to pass unexplored. The incontinence of childhood is a very common and sometimes troublesome complaint, and, when not due to any of the causes I have indicated, is probably connected with an atonic condition of the walls of the bladder, manifesting itself when the voluntary controlling muscles of micturition are temporarily in abeyance, as in sleep. In the management of these cases I place reliance chiefly upon inculcating habits of regularity in attending to such children, combined with medicinal treatment. Of the drugs in which I have the greatest confidence, I may mention belladonna and its alkaloid atropine.* I have seldom found these fail, remembering to employ them on the principle "that chronic diseases need chronic therapeutics." I am aware that a variety of mechanical means have been adapted for the treatment of this affection, such as jugums or urethral compressors, and closing the meatus with collodion, as suggested by Sir Dominic Corrigan.† Of these I think the latter is the least hurtful, and may occasionally, when other means have failed, be resorted to with advantage.

---

* I find the following formula useful in these cases:—

℞ Atropiæ, gr. i.
Acid. Acet., gtt. iv.
Alcohol,
Aquæ, āā ʒiv.

M. Four drops before each meal in a wineglassful of water.

† *Dublin Quarterly Journal*, February, 1870.

Attention to the diet is very necessary in these cases; irregularities both in eating and drinking are often attended with a condition of urine that is likely to provoke incontinence. A strictly milk diet has in some instances that have come under my notice been sufficient to effect a cure.

There is a form of irritability of the bladder which is frequently met with, especially in highly intelligent and sensitive children, at about the age of ten or eleven years, when they are entering upon the sterner forms of educational study. On examination of the urine it will be found loaded with phosphates. In remedying this condition, bromides in combination with opium will be found invaluable. The following is a formula I frequently employ for this purpose:—

> ℞ Ammon. Bromidi, ʒ i.
> Liq. Opii. Sed., ʒ ss.
> Mist. Acaciæ, ʒ i.
> Quiniæ Sulph. (neutr.), gr. xii.
> Aquæ ad., ʒ vi.
> M. One tablespoonful to be taken thrice daily.

Some slight morning aperient, such as the Hunyadi Janos water, may be required to correct the constipation the opium occasions. By the use of this prescription I have frequently seen all traces of phosphates disappear from the urine, and, with this, cessation of the irritability of the bladder. Care must be taken in these cases that the child is not submitted to an undue amount of nerve tension by reason of his educational studies.

Lastly, I would remind you that irritability of the bladder is a symptom which is by no means confined to the male sex; it is frequently met with in females.

In the same methodical manner that has already been insisted upon, the causes of the irritation must be carefully searched out, not forgetting that in females the condition of the uterus or of the rectum frequently affords a sufficient explanation. There is a cause of irritability and spasm of the bladder in females which, though uncommon, is not, I think, sufficiently recognised. I have met with some examples of it, and have found it just as satisfactory to treat as an affection somewhat analogous — namely, fissure of the anus. Ocular examination readily determines the existence of such a cause as this; the irritable appearance of a fissure at the orifice of the urethra is enough to account for the extreme sensitiveness of which the patient complains. Failing its cure by a few applications of nitrate of silver, I have found rapid dilatation of the urethra sufficient to afford relief.

We have been accustomed to regard an irritable bladder as a purely functional disturbance so far as this viscus is concerned, and for the most part it is; but we must not forget that the constant contraction of the bladder may produce changes behind it which follow as a consequence of urine-pressure—I mean dilatation of the ureters and of the kidney. I have seen this in a case of Dr. Glynn's, where there was nothing to account for these changes other than the obstacle to the escape of the urine which a constantly contracted bladder presented.

This fact is suggestive of the necessity in extreme cases of contracted bladder of maintaining for some time a means of free escape for the urine, either by the retention of a catheter, or cystotomy, in addition to the general measures I have advocated.

SIXTEENTH LECTURE.

Hypertrophy of the Prostate—Retention of Urine—Circumstances under which Retention Occurs—Treatment—Incontinence of Urine—Formation of Calculi—Operative Treatment of Enlarged Prostate.

In treating of enlargement of the prostate gland, I shall confine myself as much as possible to noticing that symptom which, as a rule, first brings the patient to ask your assistance, and, of all others, is the one most productive of distress; I mean, retention of urine.

Unlike the other urethral obstructions which have received our attention, this is a disorder which, as a hypertrophy of the gland, is only met with in advancing years, as it rarely happens to a person under fifty-five years of age.

As impeding micturition, it usually does so in one or other of the following ways:—(a) in temporarily arresting it; (b) in partially arresting it; (c) in completely arresting it. To study these conditions from a clinical point of view, let me take an ideal example, and trace it through these three stages; and I will endeavour to do so in such a way as to enable you to apply the description to actual cases you have seen in my wards, and upon which I have commented at the bedside. I do not wish it to be understood that all instances of prostatic hypertrophy in their symptoms follow the order I have taken; one or other may alone be met with, or even the last of the series of events I have sketched may be the first indication of what has gradually been taking place.

A patient of somewhat advanced age who, from time to time, has experienced various slight annoyances in connection with his urinary apparatus, after some indiscretion, such as exposure to cold or an indulgence in wine, finds himself completely unable to pass water. You are summoned urgently to him, and find, after an examination, that he is suffering from an enlargement of the prostate. You relieve him with a catheter, selecting for this purpose, not the ordinary instrument, but one considerably longer and more curved. Even in introducing this, you find a hitch as it approaches the prostate, to overcome which you introduce your finger into the rectum, for the purpose of giving it a "tilt" to help it over the obstruction. The symptom of retention is at once relieved, and possibly on the following day the patient again passes water as usual; or the retention remains, and some days, or even a week or so, may elapse before he is able to dispense with the assistance of the catheter. Now, what has happened? Clearly it is not all due to the enlargement of the prostate, or the inability to urinate would be permanent. Superadded to a prostate which is gradually enlarging, you have prostatic hyperæmia, or paralysis of the bladder, either of which conditions might be brought about by the causes to which I have alluded. If the former, the introduction of the catheter, even with the most delicate touch, is occasionally followed by more or less hæmorrhage as the instrument passes over the obstructed part of the canal. The hæmorrhage, under these circumstances, seldom does harm, and the patient is soon able to pass urine again. In cases of prostatic enlargement with hyperæmia, you usually have other indications, such as frequent and painful erections. The prostatic engorgement of advancing age has been referred to more fully in the lecture on irritability of the bladder, with which it is frequently associated. If the cause of the retention is paralysis of the bladder, due to over-distension, probably some days or even weeks may elapse before this can be overcome, for which purpose we employ, at regular intervals, catheterism. I prefer, where it can be done, passing the catheter to the retention of anything within the bladder, by which the urine can be drawn off as it collects, as cystitis is frequently produced by retained instruments. Added to catheterism, we may employ internally such remedies as strychnia, nux vomica, and iron, with the view of restoring the natural tone of the bladder.

After the patient has recovered from this temporary attack of retention, though not free from slight urinary troubles, a considerable interval may elapse before your services are again required. He may now come to you and say that he seeks your advice under very different circumstances. On the former occasion it was to relieve his retention; now, as it appears to him, it is for a very opposite condition,—he is always making water. His days are disturbed and his nights are broken by a constant desire to empty his bladder. Further than this, he complains that frequently his urine emits a very unpleasant odour. Founded upon your examination, and strengthened by the inference that if a man is constantly wanting to expel something, there must constantly be something requiring expulsion, you come to the conclusion that the bladder is now never emptied. Acting upon this, you propose to the patient to test its correctness by passing a catheter. He may possibly regard your proposal as very unnecessary, as he is disposed to think that he passes too much water rather than too little. Still, however, he submits, and you find that, though he micturated only ten minutes before your interview with him, some urine is left in the bladder. The fact is, that by the growth upwards of the prostate, the outlet of the bladder is as it were above the water level. You may possibly be able, should further proof be wanted, to convince the patient of this, by asking him, after he has made water in the usual position, to endeavour to do so with the pelvis raised—placing him on his hands and knees,—when more urine escapes The condition I have just described is well illustrated in Plate B.

Such being the state of affairs—viz., that the outlet is above the water level—what is the remedy? Clearly the remedy is a mechanical one, and we must have recourse to some artificial means of emptying the bladder, otherwise the urine, being stagnant, will decompose, and cystitis be added to the existing distress. My views as to the management of the patient under these circumstances are so much in accordance with those of Van Buren and Keyes,* that I shall quote the following paragraph:—"The question now naturally arises, Is it advisable to instruct a patient with enlarged prostate in the use of the catheter, if he has a very small amount of residuum or none at all? Most assuredly, Yes. If there is no residuum, still, with the slow advance of the disease, a time is pretty sure to come when there will be a certain quantity, or when, from the effect of cold, irritating urine, or other cause, retention may come on. It is a rule with no exceptions, that a patient with hypertrophied prostate is never safe unless he can pass a catheter for himself, any more than is a patient with hernia who does not wear a truss. Hence, in all cases, the patient should be taught the use of a soft catheter, be provided with an instrument, and instructed in the manipulation of washing out the bladder, both for purposes of cleanliness and so as to be enabled to employ medicated injections. If the amount of residuum is small, so that no material relief is afforded by the mere draining off of the urine which the patient cannot pass, still the force of the above reasoning is applicable, and the utility of washing out the bladder is equally necessary, since the liability to the formation of stone exists as well where the residuum is small as where it is large."

* *Genito-Urinary Diseases,* p. 196.

Where, then, we have residual urine, the catheter should be employed at least once in the twenty-four hours; and where there is any indication, from the smell, appearance, or reaction of the urine, that decomposition is taking place, the additional precaution of washing out the bladder should be taken. This latter operation will be further noticed in speaking of the treatment of cystitis. It is often remarkable to notice how dexterous a patient will become in the use of the catheter. Even elderly persons, when they have once gained confidence, often become most expert operators, and take considerable interest in the management of their own case, when you have clearly explained to them the object of your treatment, and they learn from experience the relief that is afforded.

We will now proceed to consider what I have taken as the third and often last phase of the disorder—namely, where retention of urine is complete. When the prostate gland has enlarged to such dimensions as to offer an obstacle—and sometimes an insuperable one—to the escape of urine from the bladder, there is no difficulty in making the diagnosis; in addition to the age of the patient who is the subject of retention, digital examination of the rectum usually reveals the state of the gland. In some instances it has attained enormous dimensions; I have seen it larger than the fist, and sufficient during life to exercise pressure upon the rectum, and so impede defæcation.

The enlargement most frequently involves that portion of the gland known, from the days of Sir

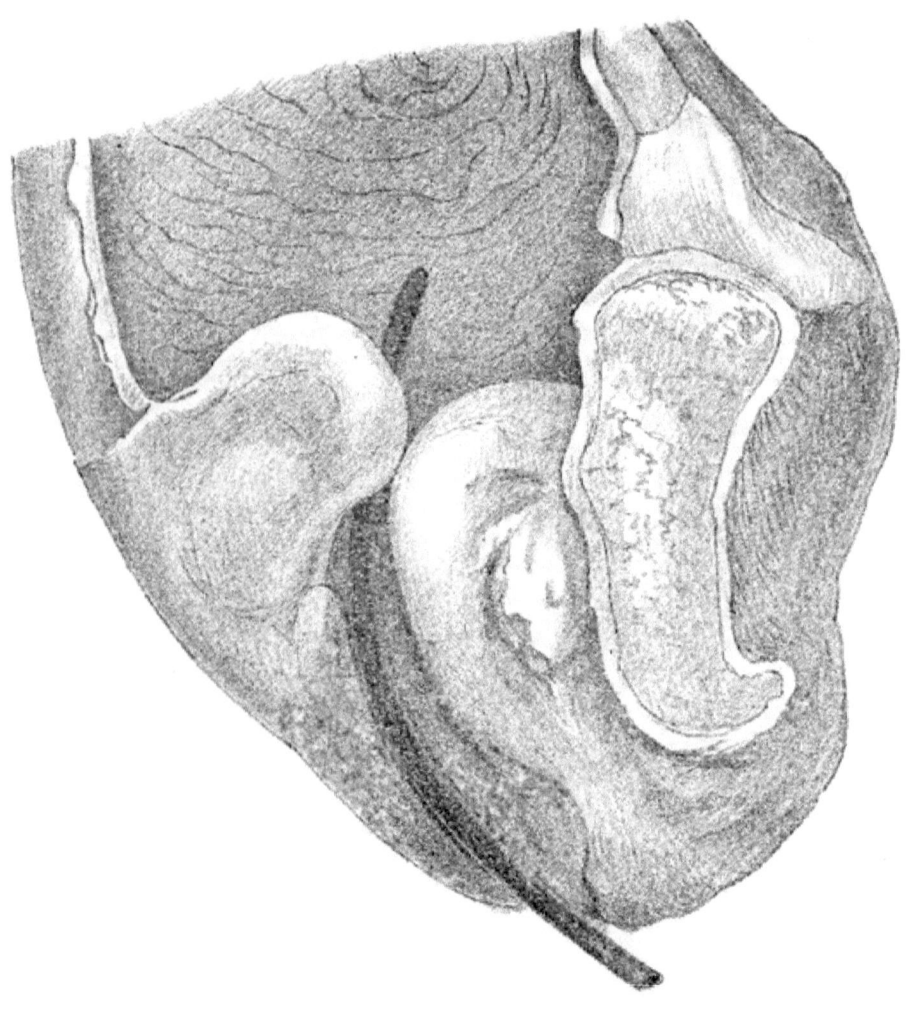

PLATE B.

Everard Home, as the third lobe, the effect of which is, as you see from Plate B, to elongate the urethra, and to obstruct the free outlet of the urine from the bladder. I mention these two points particularly, as explaining why we use a longer and differently shaped catheter when relieving retention thus caused. In the Norwich Museum there is a specimen of hypertrophy of the prostate from a patient, aged 84, who died from false passages and abscess near the gland. A catheter fourteen inches long was required, and the prostate weighed twenty ounces.*

When retention occurs, and this may happen without any premonitory symptoms, it is the duty of the surgeon to attempt catheterism. For this purpose I generally select a No. 8 gum-elastic catheter, three or four inches longer than that required for retention arising from stricture in other parts of the urethra. I again refer to the length of the instrument, as I have seen two or three instances where surgeons have failed to relieve retention, not from making false passages, but simply because their catheters have not been sufficiently long. Some surgeons prefer the French gum-elastic prostatic catheter, where the point is permanently fixed at a suitable angle. (Fig. 31.) When

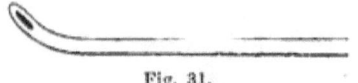

Fig. 31.

the prostate is reached, assistance may often be rendered by the finger in the rectum, in tilting the end of

* *British Medical Journal*, August 22, 1874.

the instrument over the obstruction; or, again, the expedient of passing down a stiff stylet, bent at an angle, along the catheter, is often successful in causing the point to surmount the large lobe, and so to enter the bladder.

The extent to which the bladder may be distended in cases of enlarged prostate, is often very remarkable. Several quarts of urine have been removed at a sitting, and the question as to whether the case is one of ascites or bladder-distension has arisen. In cases of largely distended bladder in enfeebled persons suffering from prostatic enlargement, it is a consideration whether the whole of the urine should be removed at once. Where the distension is great—for instance, when the collection amounts to several pints—I entirely agree in the opinion of many eminent surgeons in saying that it is better not suddenly to empty the bladder. In practice, however, in our desire to give the fullest amount of relief, we may inadvertently disregard this precaution. The objections to the removal of a large quantity of urine are these: in an enfeebled person it is apt to be followed by syncope, or, as I have seen, when the distension is thus removed from the bloodvessels of the parts, passive hæmorrhage into the bladder and out through the urethra has been the consequence. Such a loss of blood has, in this way, been fatal in the course of a few days. These effects are similarly observed after the rapid removal of fluid from other parts of the body. Syncope, after tapping for ascites, is not uncommon, and in a case I recently saw, the withdrawal of the ascitic fluid was immedi-

PLATE C.

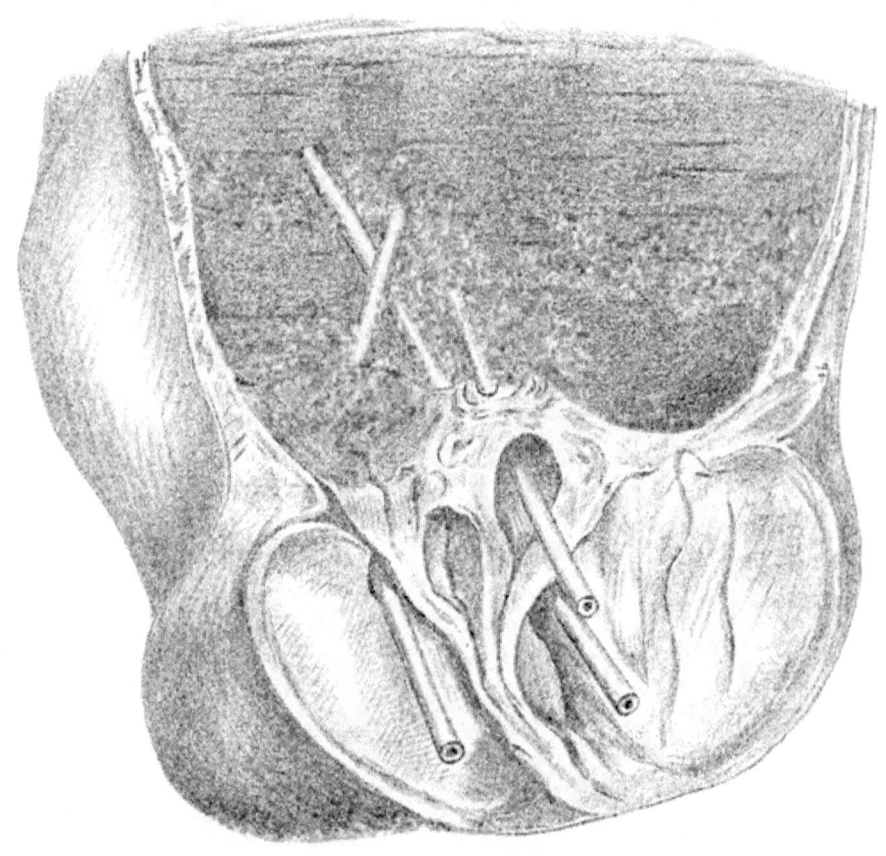

ately followed by violent hæmatemesis, which was fatal in a few hours. The removal of the tension from the abdominal bloodvessels was the only explanation of this fatality. The bladder is also more likely to regain its muscular power when it is gradually emptied than when suddenly reduced to a flaccid condition. Hence it is a good rule, in the case of a feeble person who, for some days, has been suffering from retention, and whose bladder is considerably distended, to draw off the urine by degrees.

The size and direction of the prostatic enlargement sometimes render catheterism impossible, and then the question arises as to what is best to be done. Forced catheterism, by which is meant driving the instrument through the obstruction and thus entering the bladder, has been advocated. In this specimen from the Museum (Plate C) you will see that this has been done, and the prostate completely riddled with holes, as shown by the pieces of bougie which have been introduced through several of the false passages. Though here the proceeding appears to have been, for a long time, practised with impunity, it is not to be recommended, inasmuch as, apart from the hæmorrhage and other damage it may occasion, it is calculated, from the force necessary, to set up inflammation around the gland, as I have seen in two instances, where this operation was followed by fatal pelvic cellulitis. In a case where there was this difficulty, I tapped, with the aspirator, above the pubes; the distension being removed, I was subsequently enabled to employ catheterism in the usual manner. Aspiration, as before illustrated, may be repeated almost indefinitely.

Where, from the extreme irritability of the bladder, associated with an enlarged prostate preventing catheterism, it is necessary to maintain a more permanent escape for the urine, I advise tapping the bladder above the pubes, and retaining a cannula.

Additional attention has recently been drawn to the advantages of supra-pubic puncture in the treatment of advanced forms of prostatic enlargement, by Sir Henry Thompson.* My own experience is certainly in favour of this practice, which is well referred to by an eminent writer on surgery in the last century as being " easy both to the patient and the operator." †

The objection to retaining instruments in the bladder is that they may act as causes of irritation, but as unhealthy urine is a greater source of evil, it is necessary to select the lesser of the two, remembering that the annoyance produced by a retained catheter can be reduced to a minimum by the employment of scrupulous care and cleanliness in washing out the bladder, and removing all extraneous sources of inflammation.

We must not forget that what is called "incontinence of urine" almost invariably indicates that the bladder is full—so full that it is actually overflowing,—and requiring for its relief the same treatment we employ for retention.

---

* *Proceedings of Medico Chirurgical Society*, vol. xlix.

† *The Operations of Surgery*, by Samuel Sharp, F.R.S., 1758. 7th Edition, p. 74.

PLATE D.

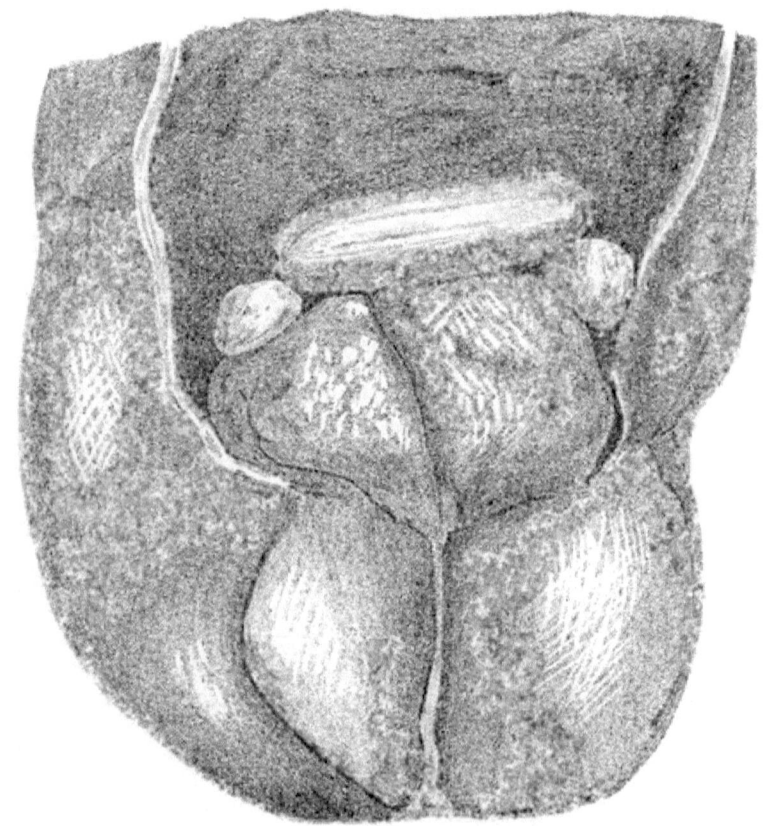

Prostatic enlargement, by inducing those changes in the urine to which allusion has already been made, sometimes leads to the formation of a calculus, the symptoms of which are obscured by the original disorder. Further than this, the hypertrophied prostate, by concealing the stone behind it, has prevented its detection by the sound, as the accompanying sketch (Plate D), from a specimen in the Museum, shows. It is a good rule, in cases of hypertrophied prostate, to take an opportunity of carefully exploring the bladder with a sound, not forgetting to sweep round the depression which is always found behind the enlarged gland. (Plate B.)

I shall, on a future occasion, have an opportunity of making some further remarks on the use of the sound in the diagnosis of stone.

My attention has hitherto been chiefly confined to the treatment of the symptoms connected with impeded micturition, which become prominent as enlargement of the prostate proceeds.

Patients, however, long before reaching the confines of three-score years and ten, some by anticipation, others by a realisation of the earlier symptoms of prostatic enlargement, not unfrequently ask advice how they may keep in abeyance the graver symptoms and complications of this affection. In advising such persons, I have been in the habit of laying stress upon the following points :—

1st. To avoid being placed in circumstances where the bladder cannot be emptied at will.

2nd. To avoid checking perspiration by exposure

to cold, and thus throwing additional work on the kidneys. In a climate such as ours, elderly persons should, both in winter and summer, wear flannel next the skin.

3rd. To be sparing of wines, or of spirits exercising a marked diuretic effect, either by their quantity or their quality. Select those which promote digestion without palpably affecting the urinary organs. A glass of hot gin-and-water, or a potent dose of sweet spirits of nitre, will not do anything towards removing the residual urine behind an enlarged prostate.

4th. To be tolerably constant in the quantity of fluids daily consumed. As we grow older our urinary organs become less capable of adapting themselves to extreme variations in excretion. Therefore it is desirable to keep to that average daily consumption of fluids which experience shows to be sufficient and necessary. How often has some festive occasion, where the average quantity of fluid daily consumed has been largely exceeded, led to the over-distension of a bladder long hovering between competency and incompetency. The retention thus caused, by suspending the power of the bladder, has been the first direct step in establishing a permanent, if not a fatal, condition of atony or paralysis of the organ.

5th. It is important that from time to time the reaction of the urine should be noted. When it becomes permanently alkaline in reaction, or is offensive to the smell, both necessity and comfort indicate the regular use of the catheter. If practicable, the patient may be instructed in the use of this instrument.

6th. Some regularity in the times of performing micturition should be inculcated. We recognise the importance of periodicity in securing a regular and healthy action of the bowels, and though the conditions are not precisely analogous, yet a corresponding advantage will be derived from carrying out the same principle in regard to micturition.

The sum of these instructions is, that as we cannot arrest the degenerative changes by which the prostate becomes an obstacle to micturition, it is of the first importance that every means should be taken to compensate for this by promoting the muscularity of the bladder, and thus prevent its being atrophied or paralysed either by accident or improper usage.

My remarks have for the most part been confined to considering how we may best palliate the inconveniences connected with a large prostate. Is there no radical measure, it may be asked, by which it may be brought down to its normal dimensions? Operative surgery, the external and internal use of iodine and iodide of potassium, astringents and sedatives in suppositories, have all been tried, but with no satisfactory or even encouraging results. Here and there are instances where surgery has, so to speak, accidentally contributed to the comfort of patients with enlarged or out-growing prostates complicated with stone, where, in addition to lithotomy, a removal of the growth, or a considerable portion of the prostate, has also been effected. In a paper by Dr. C. Williams, of Norwich, an important observation is made bearing upon this point.* In narrating the case upon which his remarks are based, it is stated—" An enlarged middle lobe of the prostate became engaged between the blades of the lithotomy forceps, and was unconsciously torn off and came away with the stone. There was free internal hæmorrhage from a deep-seated vessel, which was without much difficulty seen and secured by ligature." Further, it is said, " Three weeks later I found him in excellent health, and the wound perfectly healed. He seldom found it necessary to micturate more than once during the night." The patient, I should have added, was 72 years of age. I will further quote from Dr. Williams's remarks :—

" The removal of a large portion of the middle lobe of the prostate, though quite accidental, was attended with a happy result. The man was relieved of a trouble which, sooner or later, would have been a source of grievous annoyance to him. I witnessed the same accident in a case operated on by Mr. Cadge. In the forceps, between the stone and the blades, there came away three masses, which were apparently fibrous outgrowths of the prostate, and which weighed one drachm two scruples. In two months the wound had healed and the patient was strong and well. Mr. Cadge states: 'It has happened to me twice before to remove small fibrous tumours of the prostate gland during the operation of lithotomy, and apparently without harm to the patient.' (*Transactions of the London Pathological Society*, vol. xiii., 1862.) And

---

* Case of Lithotomy, in which an Enlarged Middle Lobe of the Prostate Gland was Accidentally Removed. *British Medical Journal*, June 15, 1878.

he gives the experience of an expert modern lithotomist on this subject, who says: ' It has occurred to me eight or ten times to bring away portions of the prostate, and without noticeable injury to the patient. In more than one instance it was the prominent third lobe which got between the handles, anterior to the hinge, and was torn off entire; I have never known unpleasant results to the patient, and sometimes he has been benefited in after-life, by having got rid of an useless impediment to a natural function. I would not willingly that such an occurrence should happen, and I try to avoid it by turning the blade of the forceps to the lower angle of the wound as I leave the bladder; but when it does occur I lay no account by it.'"

Such facts as these have an important bearing upon the operative treatment of enlarged prostate. Mercier's practice of internal incision of the prostate cannot be said at present to have a place in the operative surgery of the urinary organs. The claims of this proceeding to consideration have recently been advocated by Dr. Gouley, of New York.*

Ergot and ergotine, administered by the mouth, and subcutaneously, have been vaunted as possessing the power, not only of preventing further enlargement of the gland, but of diminishing its size. Dr. Atlee† has found considerable advantage follow the use of this drug, and his experience is, to some extent, corroborated by other practitioners.‡ My own practice is

* *The Lancet*, July 17, 1880.
† *New Orleans Medical and Surgical Journal*, August, 1878.
‡ *British Medical Journal*, September 28, 1878.

very favourable to the use of ergot in cases of difficult micturition in connection with enlarged prostate; but my conclusion is, that the gain comes from the stimulating effects of the ergot upon the muscular coat of the bladder, rather than from any diminution it effects in the size of the obstructing gland. I fear the curative treatment of enlarged prostate must yet await the result of further pathological enquiry. Whether it is determined by certain states of urine, by alterations in the vascular supply, such as a permanent varicose condition of the prostatic veins, or disturbed nerve control, are problems which, amongst others, remain to be solved: in the meantime we must be contented with those palliative measures which, in their successful application, are capable of contributing much to the longevity and comfort of those suffering from this inconvenience.

SEVENTEENTH LECTURE.

INFLAMMATION OF THE BLADDER—ATONY.

CYSTITIS, or inflammation of the bladder, is an affection which you frequently have an opportunity of seeing in the surgical wards. It rarely occurs as an idiopathic disorder, but commonly in connexion with some other derangement of the urinary system. It is to the latter aspect that I shall make more special reference; and in doing so it will be my endeavour to classify the circumstances under which it arises.

First: As produced by the extension of inflammation from some other part, as in gonorrhœa. This may be called metastatic cystitis.

Second: As a consequence of obstructed micturition, as in stricture, or hypertrophy of the prostate—the cystitis of obstruction.

Third: As produced by an irritant in the bladder, such as a calculus or a growth—the cystitis of direct irritation.

It is not uncommon to find cystitis, in various degrees, occurring as a consequence of gonorrhœal urethritis, it being generally considered as an extension of the inflammation to the mucous membrane of the bladder along the urethra; like other metastatic inflammations, I have noted that the primary disorder often abates as the change in locality takes place, as if the force of the action were concentrated on one spot.

In the slighter forms of cystitis resulting from gonorrhœa, or when provoked by such causes as exposure to cold, we have this condition indicated by frequent micturition, and urine more or less loaded with mucus. In the severer forms, in addition to constitutional fever, the bladder is intolerant of the presence of urine within it, as indicated by the extreme frequency of micturition, and the distress and tenesmus produced by the contractile power necessary to expel it. The urine becomes purulent, and a discharge of blood not unfrequently terminates the act of its expulsion.

I have noted the greater liability to cystitis and bladder irritation, as a complication of gonorrhœa, at seasons of the year when sudden changes in temperature are apt to occur. Hence the importance of providing against this by suitable clothing and avoiding exposure to keen winds.

The treatment of this kind of cystitis must be in correspondence with the degree of inflammatory action that is taking place. In the milder forms, where the term "irritation" best describes the extent to which the bladder is implicated, the suspension of any kind of abortive local treatment, so far as the gonorrhœal discharge is concerned, is at once necessary. Rest, in the recumbent position, and soothing applications in the form of hot opiate fomentations and sedative suppositories* must be substituted. Of all the demulcents I have been in the habit of prescribing, I find the decoction of the ulmus fulva, or slippery elm, in combination with the succus hyoscyami, affords the speediest relief.

In the acute form of cystitis antiphlogistic means will, in addition to the above-mentioned remedies, be required. Nothing soothes the patient who is suffering from an acutely inflamed bladder more than the local abstraction of blood. A few leeches applied to the perinæum speedily removes that dreadful feeling of tension about the neck of the bladder which is always more or less complained of; as the acuteness of the attack abates, buchu and uva ursi are usually administered with the best results.†

In the treatment of cystitis reference is often made to the reaction of the urine as indicating the necessity of administering either acids or alkalis. On this point I may say that our object should be to obtain that condition of the excretion which most nearly corresponds with its normal state, as being the least likely to provoke irritation. I mention this here again, as we sometimes find that alkalis are poured in with a vigorous hand, and quite regardless of the fact that

---

* Anodyne suppositories.
℞ Morphiæ Mur. g. ½.
Pulv. Ipecac. g. ¼.
Ol. Theobrom. q. s.
Make a suppository—to be introduced every three or four hours.

The following combination is exceedingly useful :—
℞ Infus. Uvæ Ursi, ʒiss.
Infus. Lupuli, ʒi.
M   The draught to be taken every three hours.

healthy urine almost invariably has an acid reaction. In cases of gonorrhœal cystitis, where the disorder has a tendency to become chronic, frequency in micturition and purulent urine remaining after the other acute symptoms have subsided, I have found benefit from the use of copaiba, or the oil of sandal wood; these remedies are, however, not well borne where there is general febrile disturbance.

The *second* form of cystitis, or that following obstruction, is caused by the decomposition of retained urine, and consequently occurs in cases of organic stricture of the urethra, of enlarged prostate, and of paralysis where urine is allowed to remain in the bladder, and, by chemical change, to become offensive. As the treatment of cystitis occurring under these circumstances resolves itself into the removal of the cause of the obstruction and the mitigation of the consequences so produced, it will be necessary for me here to refer to those mechanical means for the local treatment of the bladder complication. These include the regular emptying of the bladder with the catheter and its thorough cleansing by water or suitable medicinal agents. For the latter purpose you see me usually employ a glass funnel to which is fitted about two feet of india-rubber tubing and a gum-elastic catheter. On the catheter being introduced into the bladder it is connected to the funnel by means of the tubing. When the funnel is elevated and water poured in, the latter, by hydrostatic pressure proportionate to the calibre and length of the tubing, is forced into the bladder; by lowering the funnel below the level of the patient, the water escapes from the bladder. In this way the viscus can be thoroughly cleansed in a very few minutes, and with little inconvenience to the patient. Some surgeons use a double catheter, to which is fitted a gum-elastic ball or syringe; I prefer the

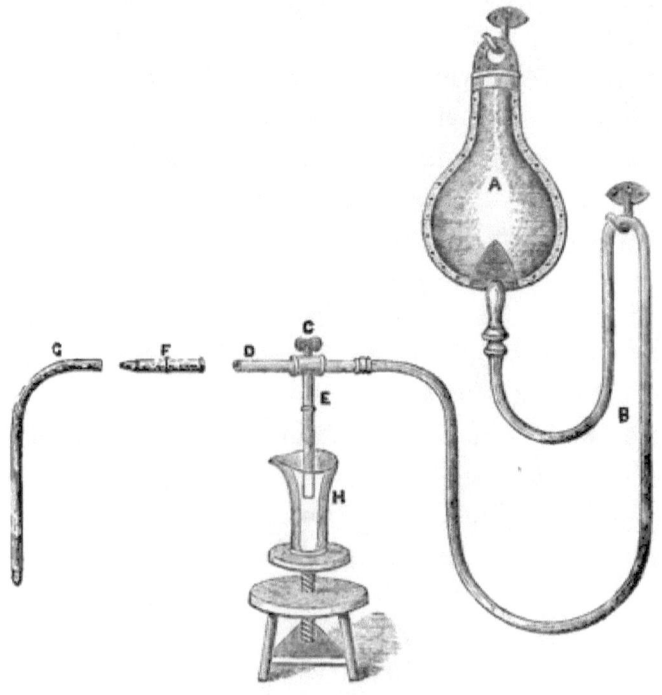

Fig. 32.

former plan, one which I find generally accords with the feelings of patients who have made a trial of both.

Where it is necessary to perform this operation for some time, a modification of the funnel arrangement may be recommended. My attention was first called to this contrivance by my friend, Dr. Fifield, of Boston, and, as it is little known, a brief description of it may be useful. From the accompanying sketch (Fig. 32) it will be seen that the apparatus consists of a vulcanised india-rubber bottle (A), capable of holding a pint of fluid, which, by means of a ring, can be suspended to any convenient hook; a piece of tubing, five feet in length (B), terminating in a stop-cock (c), which permits fluid to flow either through the catheter end (D), or the outlet pipe (E), according to the direction in which the tap is turned. A conical metallic catheter mouth-piece (F) completes the connection between the catheter (G). A soft rubber catheter is generally preferred. The instrument is used in the following way :—The bag, being filled with the fluid to be injected, is hung up about six feet from the floor. The stop-cock (c) is then turned until some of the fluid escapes, so that no air is allowed to enter the bladder. The patient, being in the erect position, then introduces the catheter and connects it with the tubing. By the alternate action of the tap (c) the fluid is made either to enter the bladder or to escape; if the latter, it passes into the receptacle (H). The instrument can be readily adapted to the recumbent position. Beyond other advantages I have found the apparatus to possess, it enables patients to perform this operation without assistance.

In cases where there is hæmorrhage from the bladder, and the catheter soon becomes blocked with clots, I have on several occasions found Clover's apparatus for the removal of detritus after lithotrity of much service. This consists of a large-eyed catheter (with a bougie for a stylet), to which is fitted a gum-elastic bottle capable of injecting fluid with as much force as it would be, under any circumstances, desirable to use. In washing out the bladder take care not to distress the patient by the manipulation employed. The injections should not exceed four ounces in quantity, and they should not be too often repeated.

Care should be taken to prevent the forcing of air into the bladder along with the fluid to be injected. When the urine is bloody, the presence of air in the bladder speedily leads to putrefactive changes and the evolution of gas in considerable quantities, which not only is most offensive, but is capable of producing retention of a very painful character. I have observed these effects to follow, and hence the necessity for this caution.

There is one source of annoyance which patients occasionally complain of, and which I will mention, as it is quite easy to avoid, and you may not at first think of it. As the bladder is expelling the last portion of the injection, if the surgeon is holding the catheter, he sometimes feels a slight click or shock, which the patient with a sensitive bladder is conscious of and rather dreads, though it is very momentary, on the next occasion the operation has to be repeated. I believe it is caused by the mucous membrane being sucked into the eye of the catheter as the bladder is emptied of the last few drops. It is to be avoided by carefully watching the flow, and withdrawing the catheter until the end is well within the prostate as the last portion of fluid escapes. It is better to do this than to obviate the inconvenience by having a catheter with more than one eye, as such an instrument is liable to break. A catheter for this purpose should have the eye of moderate size, with its edge bevelled, like the American instrument, so as not to scratch the urethra; the eye should be close to the end of the instrument, to avoid an unnecessary length of catheter being introduced into the bladder.

For bladder injections, for cleansing purposes, tepid water with about five grains of borax to the ounce, is generally employed; where the urine remains in an unhealthy condition for some time, a few drops of nitric acid or of the tincture of perchloride of iron may be added with advantage.

In cases of muco-purulent urine following cystitis, consequent on stricture or stone, I have recently been using with great benefit injections of quinine. My attention was first called to the value of quinine for this purpose by a paper by Mr. Nunn,* who speaks of its action as a bactericide in that chronic form of cystitis known as catarrh; when the urine is purulent and offensive, I have certainly found it to be exceedingly efficacious. The neutral sulphate of quinine dissolved in distilled water, in the proportion of two grains to the ounce, will be the most suitable. If the solution is not quite clear, a drop of dilute muriatic acid should be added, and then it should be strained. A quantity, not exceeding three ounces, of the solution may then be injected by a catheter and india-rubber bottle,

* *The Lancet,* February 23, 1878.

the patient being instructed, if possible, to retain the injection until it is spontaneously expelled. I have found the internal administration of quinine, in ten-grain doses, not only act as a sedative to the bladder, but useful in arresting putrefactive changes in the urine. Its efficacy for this purpose has been urged by Dr. Simmons, who, in explaining the nature of this action, refers to an observation by Dr. Kerner, that seventy per cent. of the drug is eliminated by the kidneys in from three to twenty-four hours after it has been taken.*

To alleviate the extreme irritability of the bladder, which often remains after the more active symptoms of inflammation have passed away, a solution of morphia, injected into the bladder with a gum-elastic catheter to which a ball syringe is attached, often gives the patient a good night after rectum suppositories in various forms have been tried.

For a similar object, I am now employing in these cases vesical suppositories, containing morphia, belladonna, bismuth, and other soothing agents. I put these into the bladder by means of a pessary-catheter, which has been made for me by Messrs. Krohne and Sesemann. The instrument consists of a silver catheter, open at the end, in which the pessary is placed. By means of this instrument the whole of the urine is first drawn off, after which, by pressing the stylet, the pessary is projected into the bladder. The pessaries are made of the oleum theobromæ, and are so shaped as to fit in the open end of the catheter, thus giving it the

* *American Journal of the Medical Sciences*, April, 1879.

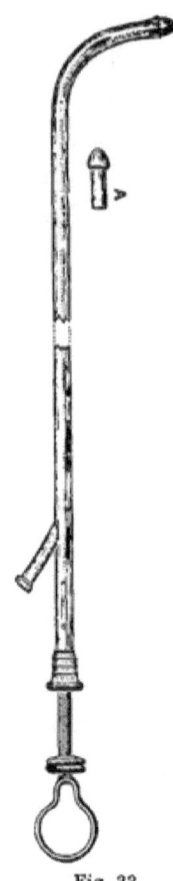

Fig. 33.

appearance of an ordinary instrument, and facilitating its passage into the bladder. The shape of the pessaries is shown in the sketch (Figure A); they contain various medicinal applications.* A grain of morphia, introduced into the bladder in this way, and repeated twice in the twenty-four hours, has, in several instances, completely and permanently relieved the most distressing symptoms of irritation. I have extended the application of these vesical pessaries to other cases where astringent and direct applications to the bladder are indicated. I have seen much benefit follow the use of these pessaries, and feel sure that a still more extended experience will show the advantage, under certain circumstances, of thus being able to medicate the bladder. I would like to add to my own testimony that of a member of our profession who, after the experience gained by an illness extending over twelve years, writes me from a considerable distance: "I have received more benefit from the pessaries of cantharides than from all the remedies I have tried during the twelve years I have suffered from retention."

Of internal remedies used in the treatment of the chronic form of cystitis there are two which I would

* These pessaries are made for me by Messrs. Symes & Co., Hardman-street, Liverpool.

specially mention. In some of these cases, where other means have failed, I have occasionally been struck with the extreme rapidity with which pus disappeared from the urine under the influence of chlorate of potash. It is best prescribed in the proportion of half an ounce in a pint of water, a tablespoonful being taken every two or three hours, or with quinine as previously stated. Again, in other cases of this description turpentine, combined with a little mucilage, is often of great value; five minims should be taken two or three times a day.

In regard to diet, Dr. Johnson has shown the value of milk exclusively, both in acute and chronic cases of cystitis. The truth of this observation I have frequently noticed, the effect of the milk being to render the urine less irritating. Many people cannot digest milk unless it is first cooked—that is to say, boiled and allowed to become tepid or cold, according to taste. When, even in this way, it disagrees, it may be peptonized, as recommended by Dr. Roberts.*

Atony is a not infrequent consequence of retention of urine, and is often seen under the circumstances I am now describing. From long-continued over distension, the bladder becomes a mere flaccid receptacle, and loses all power of expelling its contents. To

---

* To Peptonize Milk.—Fresh milk is diluted with water in the proportion of three parts of milk to one part of water. A pint of this mixture is heated to boiling, and then poured into a covered jug. When it has cooled down to between 140° and 150° F., three teaspoonfuls (ʒiii.) of the Liquor Pancreaticus (Benger's), and 20 grains of bicarbonate of soda (in solution), are mixed therewith. The jug is then placed under a "cosey" in a warm situation for one hour. At the end of this time the product is again boiled for a couple of minutes. It can then be used like ordinary milk.

prevent this becoming permanent, much may be done by the mechanical measures I have been advocating. In addition to these, medicines, such as iron, nux vomica, and strychnia, have been administered with varying success.

Dr. Glynn speaks very highly of the tincture of cantharides, given in twenty-minim doses, in cases of paralysis of the bladder arising from affections of the spinal cord; it is an old-fashioned remedy, which, in addition to its diuretic properties, probably exercises, as Dr. Glynn suggests, a direct stimulating action upon the bladder by its presence in the urine.

Professor Von Langenbeck has recently, I see, found considerable benefit in these cases from the use of hypodermic injections of ergotine. "In all the cases (three) there was an immediate increase in the contractile power of the bladder, so that the patients passed more water than they had previously done; and, after a few days, the bladder had so far recovered its force that scarcely any urine remained in it after micturition."* In two cases where ergotine has thus been tried, at my suggestion, the patients certainly appeared to gain muscular power; both of them suffered from impeded micturition, with considerable prostatic enlargement.

The last form of cystitis I have here to notice, is that produced by an irritant within the bladder itself, such as a calculus or a tumour—causes which may or may not permit of removal. These, however, will be more conveniently considered later on in con-

---

* *Medical Times and Gazette*, April 7th, 1877.

nection with the other symptoms, of which this may be only one.

I need hardly remind you that there are causes of cystitis other than those I have enumerated, for the sake of convenience, in a tabular form. A paralysed bladder, as we see in disease and injury of the spinal cord, is, sooner or later, almost sure to become an inflamed one in the way that has already been explained. In the surgical wards we find this in cases of fractured spine, where there is retention. Catheterism and washing out the bladder will do much towards mitigating the distress of the patient and averting a fatal issue; for where recovery has followed — as in instances recorded by Mr. Manifold[*] — much of the success is due, I believe, to the absence of inflammation of the bladder. In employing catheterism in these cases, we ought not to forget that, owing to the absence of sensibility in the parts, much damage may be inflicted by an injudicious employment of instruments, without the patient expressing that consciousness of pain which otherwise he would do. The greatest care should consequently be exercised, in drawing off the urine, to avoid any laceration of the urethra or bladder, which, considering the state of the urine, would be sure to provoke further complications. Almost the whole comfort of the patient suffering from fracture of the spine depends upon the manner in which his urinary symptoms are anticipated and managed.

In the treatment of cystitis as it occurs in

[*] *Liverpool Medical and Surgical Reports*, vol. ii.

females, you can have no better instructions than those contained in a very practical paper on this subject by Dr. J. Braxton Hicks. The author points out how little is to be expected from internal remedies beyond correcting such abnormalities as the diseased state of the urine and disorder of the functions generally, and how much may be done by local treatment. Reliance is chiefly placed upon washing out the bladder with slightly acidulated warm water until it is clear of phosphates and mucus, and afterwards injecting, with a view of its retention, a solution of morphia. Subsequently permanganate or chlorate of potash is employed in a similar manner. On the subsidence of the acuter symptoms, injections of tannin or of perchloride of iron, followed by morphia, are substituted, and are again changed as the bladder becomes less sensitive for more potent astringents, such as the nitrate of silver. "The benefit of such management is very marked in cases of paralysis where, from retention or the rapid ammoniacal decomposition of the urine, the distress and constitutional irritation are very distressing; and thus we can often lessen the chance of the extension of the irritation to the kidneys. Again, in malignant disease, the simple injection of acidulated warm water gives amazing comfort, removing the phosphates and ammonia, and when to this is added the morphia, a wonderful comfort is felt. Indeed, so much relief is obtained that, with a large calculus in the bladder, its presence is almost entirely unfelt if morphia be daily injected."[†]

[†] *British Medical Journal*, July 11th, 1874.

EIGHTEENTH LECTURE.

On the Formation and Physical Constitution of Urinary Calculi.

Such an enquiry is an essential preliminary to considering the prevention and treatment of stone.

To be able to define the earliest condition favouring the formation of calculus would be suggestive of the line of treatment necessary to prevent its occurrence. We are familiar with the chemical composition of the various deposits the urine is capable of yielding; we meet with individuals who go on year after year passing urates, uric acid, or oxalates, and yet remain free from the suspicion of stone, these salts making their way through the urinary passages just as readily as flour will pass through the minute holes of a dredger. Why do they not concrete? What is the law or condition that determines their concretion? Is the human body capable of favouring a process which is observed in the inorganic world, where masses varying in size are built up by crystallization. An examination of calculi reveals facts opposed to such an idea as this, as by far the greater majority evidently consist of mixtures of concreted salts put together in layers without any resemblance to crystallization. Are these calculi, then, formed by what may be described as a cementing process where the inorganic particles are massed together by some organized cement, such as albumen, mucus, or other similar element which the blood or the tissues are capable of providing?

The objection to this, which I would speak of as a cementing process, is the apparent absence of any provision for the hardening of the cementing material. The constant presence of moving urine would in itself prove an obstacle to an explanation based upon such premisses, and as forcibly so as it would in any other analogous process where, by means of a gum or a cement, an endeavour was made to mass together particles of sand or other similar inorganic material whilst exposed to the action of an uninterrupted current of water. Something more than a simple cementing process must be looked for.

In the case of the triple phosphates, the process of stone formation appears to me to more closely resemble what is known as precipitation, these salts being abundantly deposited when the urine is rendered alkaline by the decomposition of its urea, as we see in cases of residual urine and chronic cystitis.

Illustrations of the precipitation of phosphates we have in various morbid conditions of the urinary apparatus. We see it where foreign bodies, such as catheters or bougies, remain for some time in the bladder; in intra-cystic growths, as villous tumours, the fringes of which become coated with triple phosphate; and lastly, in the casing in of other calculi with this deposit.

And the latter circumstance has a special interest,

as it shows how one morbid process may be superadded to another for the purpose of rendering inert the effects of the one preceding it.

Take, for instance, the calculus the section of which I give as an illustration. The central portion of this calculus consists of oxalate of lime of most irregular outline. So long as the urine remained acid it appears to have grown by successive additions of that material of which it is so largely and centrally composed. When, however, by reason of its angular and uneven form it had provoked vesical irritation and cystitis, the urine became alkaline, and phosphates were precipitated, by which the uneven stone was coated over until all its angles and irregularities were obliterated, just as completely as if it had been done by a mould of plaster of Paris. It would then seem, in its altered and even form, to have ceased to be a source of vesical irritation and of alkaline urine, for we notice the reappearance of the original deposit on the margin of the phosphates, indicating that the urine again became characteristic of the diathesis which in the first instance led to the formation of a stone. Surely in this we may recognize a distinctly conservative process; and this consideration explains the formation of certain forms of alternating calculi of phosphates and other salts. In these the irregularities of the primary formation are covered in by the phosphates, until the calculus is rendered smooth; then, with the urine again becoming acid, we

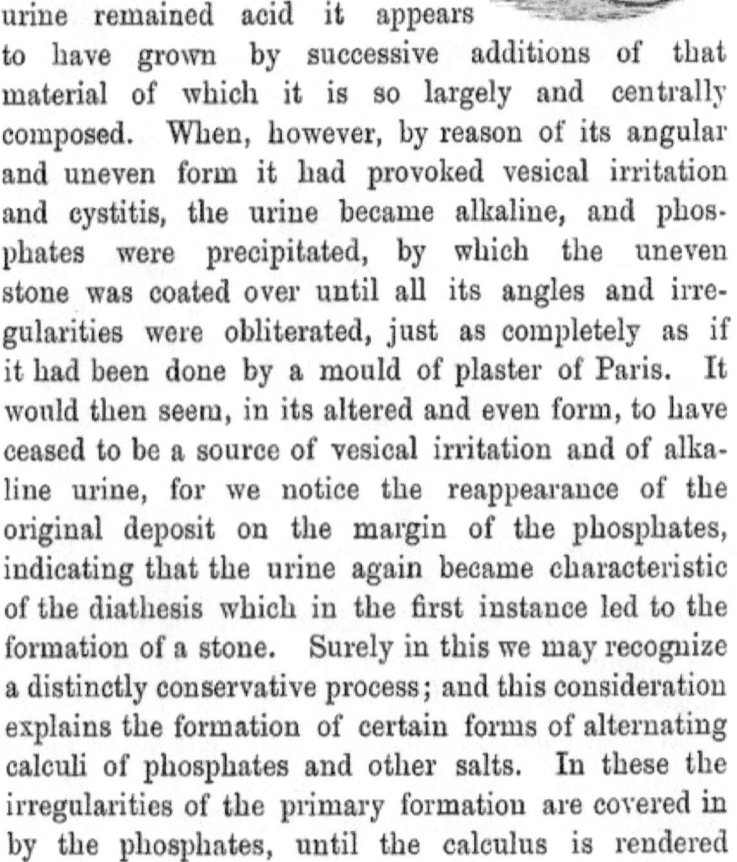

Q

have the reappearance of the original salt on the periphery of the calculus, until this in its turn requires smoothing down by the phosphates provoked by cystitis, and *da capo*. The clinical history of the patient from whom the stone was taken that has served to illustrate these remarks seemed to warrant the assumption I have drawn as to the probable sequence of events; and as botanists and geologists are enabled to explain much of the history and age of vegetable and strata from an examination of their sections, so the pathologist, by a similar process applied to calculi, can describe many of the clinical circumstances which characterized their development and increase; how, at one time, they progressed unostentatiously by an aggregation of that material which represented a constitutional disorder or a diathesis, whilst at another their appearance and the massing of phosphates upon them speak of intense vesical irritation, alkaline and offensive urine, and such-like local distress.

But though precipitation, as seen with the phosphates, may be the first step in the process of stone formation, it cannot be the only one; for by what forces or actions are the separate particles of the precipitate drawn or held together? Can what is seen to follow the lodgment of a piece of bougie or other foreign substance in the bladder be imitated out of it?

We must look in directions other than those I have indicated for an explanation as to how the greater majority of calculi are formed, and this, I think, will be found in something more nearly approaching a physiological than a purely physical process.

By the observations of Rainey[*] and of Ord[†] it has been shown that some salts, in the presence of a colloid material such as gum or albumen, yield, not crystals, but certain bodies to which Carter[‡] has applied the term "submorphous," having the peculiarity of adhering not only to existing surfaces, but also to each other, in laminar series. In the urine may constantly be observed urates presenting an appearance precisely similar to these submorphous forms, and the existence of an organized material partaking of the nature of a colloid has been demonstrated as existing in a very large proportion of urinary calculi. In reference, then, to the concurrence of these two events, which are necessary to the production of a calculus by molecular coalescence, I will quote a paragraph from the work by Dr. Vandyke Carter, to which I have already referred :—" Regarding, first, the probabilities of the case, it seems to me that the necessary conditions for the operation of molecular coalescence may at times well occur in the living human subject ; thus an excess of mucus, perhaps altered in character in the urinary passages, or the effusion of albumen, fibrine, or blood and the like, say from congestion of the kidneys or from irritation of the urinary tract, would furnish a colloid medium, with which uric acid,

[*] "Precise Directions for the Making of Artificial Calculi, with some Observations on Molecular Coalescence," by G. Rainey.—*Trans. Microscopical Society of London*, vol. vi., 1858.

[†] "On Molecular Coalescence, and on the Influence exercised by Colloids upon the Forms of Inorganic Matter," by Dr. Ord.—*Quarterly Journal Micros. Science*, vol. xii., New Series, 1872.

[‡] *The Microscopic Structure and Formation of Urinary Calculi*, by Dr. H. Vandyke Carter.

the urates, or oxalates themselves, perhaps in excess, could combine in the manner before described." Added to this is a note from Rindfleisch (vol. ii., p. 143) : " I have long been in favour of the view that the epithelial cells with which the straight tubes are lined generate a colloid material in their protoplasm, which they pour out into the interior of the tubes."

The view as to the determining action of colloid material in the formation of stone has, I take it, received a practical corroboration in certain facts which have been advanced in the course of enquiries having for their object the etiology of calculous disease.

Amongst the most important of these is the Address on Surgery, by Mr. Cadge, delivered at Norwich on the occasion of the meeting of the British Medical Association in 1874. Taking the commonest variety of stone—namely, uric acid— I do not think that Mr. Cadge has proved that the people of the stone-forming district of which Norwich is the centre, actually secrete larger quantities of uric acid than others; proof could readily be advanced if this were so. The explanation of the frequency of stone in this locality lies rather in the circumstances which determine the aggregation of the particles of which stone is composed. From the evidence before us, it appears to me that the prevalence of stone in a district bears a corresponding ratio to that of the hardness of water. " Hard waters," as Prout remarks, " have a great influence in producing stone," and though this is generally admitted, and chemistry has been most zealously applied, it has failed to afford us any satisfactory clue. I submit that the explanation of this is physical rather than chemical or physiological, and lies in the fact that the action of certain waters, denominated hard, occasions direct irritation of the urinary passages; that this necessarily increases the quantity, or alters the character, of the mucus secreted in the urinary tract, and that in this way is furnished the colloid requisite for the aggregation of the normal inorganic constituents of the urine by molecular coalescence.

In reference to the use of milk as an article of diet, Mr. Cadge supplies us with evidence strongly corroborative of the view I am taking, for he tells us " that the prevalence of stone amongst the children of the poor is largely due to their not obtaining a proper and sufficient supply of sound milk;" and, further, that the abundance of stone in children "will be found in strict accordance with the difficulty of procuring milk."

The soothing effect of milk in a variety of forms of irritability of the urinary passages has been pointed out by Dr. George Johnson, and is now generally admitted, inasmuch as many cases of this kind are completely remedied by the employment of a milk diet.

Now, what is the effect of substituting for the natural food of children a diet in which, amongst other things as drink, water of an intensely hard character must be largely consumed without the soothing influence of a milk diet being combined? I do not think I am begging the question in saying that amongst other changes the most striking will be a large increase in the proportion and viscidity of the urinary mucus. Variations in the quantity of mucus following changes in articles of drink I have frequently had occasion to observe. The truth of this can be readily tested. The non-occurrence of stone in other places where also the water is hard may be explained by the probability of any irritation caused to the urinary organs being compensated by the freedom with which the children can obtain milk as an article of diet. In Norfolk it is not so; there not only is the water exceptionally hard, but the difficulty of obtaining milk, by which its injurious effects on the urinary organs might be counteracted, in the case of the children of the poorer classes, who chiefly suffer from stone, is described by Mr. Cadge as "lamentable."

And that which is observable as following the irritating influence of certain hard waters in producing an excess of mucus in the urine, is equally noticeable in other forms of irritation produced by altered characters of the urine: the excess of uric acid in the urine of the gouty; of the oxalates in the dyspeptic; of the phosphates in the unhealthy urine of those suffering from enlargement of the prostate, are analogous causes, equally capable of producing that excess of mucus which is a natural consequence of the drinking of water unfitted for human consumption. I have frequently noted the excess of mucus in the urine of persons who are passing quantities of certain crystals of uric acid. In these individuals the mucus is sufficient to provide the colloid necessary to molecular coalescence of the inorganic particles, and consequently such subjects must be regarded as extremely liable to stone formations both in the kidneys and the bladder. These persons should always be put on their guard.

The practical value of the theory of stone formation by molecular coalescence, to which the presence of a material capable of playing the part of a colloid is essential, lies in this—in the prevention of the necessary conditions being provided, or if this is inevitable, the rendering of them unsuitable for promoting those changes which ultimately result in the formation of a stone; and this I submit we are capable of doing to a very considerable extent. If the process of stone formation could be explained only by altered proportions of the inorganic constituents of the urine, would not pathological chemistry, considering how searching the enquiries have been, have done something to determine it? or should we not expect to find some peculiarities in the urine of those residing in a particular district prone to such formations? But it is not so; we must therefore content ourselves with searching for circumstances favouring the aggregation and consolidation of the natural constituents of the urine; and I take it this is to be found in the application of the theory of molecular coalescence through colloids which has been advanced chiefly by Rainey and Ord.

NINETEENTH LECTURE.

Spontaneous Fracture of Calculi — Varieties in their Composition and Shape.

The formation of calculi may advantageously be followed by some observations on what I may call their self-destruction. Such a consideration is not without its object in a practical treatise, for could it be shown that such effects are due, not to accidents, as falls or blows, but to certain physical or chemical changes brought about by the constitution of the calculus or the conditions with which it was surrounded, we might provide ourselves with data which would be of material assistance in further enquiries having for their object the treatment of calculi by solvent agencies.

Cases of spontaneous fracture have been recorded by Dr. Ord, the late Mr. Southam, and other observers.[*] In these I do not include instances where a stone in the bladder has been fractured accidentally by a fall or a concussion, as by jumping from a carriage; I allude to cases where there are good reasons to believe that other agencies have been at work. Such, for instance, as the case stated by Dr. Ord, to which he adds the remark, "I maintain, therefore, that the fresh specimens support the idea which I advanced upon examination

[*] *British Medical Journal,* Jan. 4, 1868.

of the first: that in a changed state of urine the nucleus had become swollen, and acted as a bursting charge in a shell."*

A case of this kind recently occurred in my own practice. A gentleman, aged about 60, with an enlarged prostate and a slight stricture, arising from injury to the urethra, was under my care for some difficulty in micturition. He was also seen with me by my colleague, Mr. Banks, and we both considered that it was exceedingly probable he was suffering from stone; but, as nothing could be done for this complication until the urethra was dilated, sounding was postponed. The patient was so much better for the treatment his stricture received that he contented himself with occasionally passing a bougie, and failed to comply with our wish to have his bladder explored for stone.

Some months afterwards he called to say that our view was correct; that after some trouble he had passed a stone in several fragments, and was now quite free from all urinary inconvenience.

He brought me two of the fragments he had passed, merely as specimens, having, much to my regret, thrown the other portions away. I examined these pieces, and found them to be fragments of a calculus of some size, which I was of opinion, from the appearance presented by the pieces, had undergone spontaneous fracture, and thus escaped from the bladder. These portions are preserved in the School of Medicine Museum, and careful examination of them by means of a magnifying glass reveals the existence of spaces

* *British Medical Journal*, Sept. 7, 1878.

between the laminæ. I cannot comprehend how these fragments thus became broken up other than by forces acting from within.

In a paper on this subject read before the Pathological Society, Dr. Ord concludes that there are three ways in which a calculus may spontaneously fracture. (1) From forces arising within the calculus itself; (2) from molecular disintegration; (3) from weakness of the layers within the crust allowing of its fracture.*

In reference to the second mode of spontaneous disintegration, I would remark that it is probably effected by some altered relation in the colloid to the salt, by which cessation of cohesion, erosion, or falling to pieces is brought about.†

I cannot help thinking that a distinguished predecessor of mine on the medical side of the Royal Infirmary—I allude to Dr. Matthew Dobson, F.R.S.—founded some views in a work entitled, "A Medical Commentary on Fixed Air," published, just one hundred years ago, on observations derived from a knowledge of the spontaneous fracture of calculi under the circumstances to which I have referred. In the course of his remarks he speaks of changes observable in calculi exposed to the action of altered states of the urine, which have been brought about by the use of waters containing fixed air, similar in character, I take it, to certain German spa waters. The effects upon

---

* *British Medical Journal*, May, 10, 1879.

† "Mr. Rainey has made known observations on the dissipation of the submorphous globules composed of lime carbonate, which claim notice in this place because of the light they throw upon the possible disintegration of calculi within the bladder."—Dr. H. Vandyke Carter, *Op. cit.*

the calculi so treated appear to me to be in imitation of those observed in connection with their spontaneous fracture, rather than as arising from the action of solvents. In whatever light it is taken, the communication is one of considerable interest, as bearing upon the views more recently advanced on the treatment of calculi by lithontriptics, and consequently is worthy of a place in the literature of this subject.

Time will not permit me to enter further into the enquiry. I must therefore content myself with this brief allusion to the importance of studying spontaneous changes in calculi, in view of the light they throw on the treatment of stone by solvent and chemical agencies.

In discussing such a subject as the formation and disintegration of urinary calculi, we cannot fail to recognize the application of microscopical and collateral investigations to practical surgery, the full force of which can only be appreciated by a perusal of the philosophical papers to which reference has been made, and to which I fear I have scarcely done justice in my endeavour to adapt the views contained therein to clinical purposes.

Of the chemical composition of calculi I shall not say much. With some rare exceptions they may be said to consist of urates, oxalates, and phosphates, and I have given them in the order of their relative frequency.

As representing the stone formation in this neighbourhood, Mr. Paul has made an examination of the human urinary calculi in the Museum of the Liverpool

Royal Infirmary School of Medicine, which includes one hundred and sixty specimens, with the following results :—Of the *simple* calculi, forty-nine consisted entirely of uric acid or urates, thirteen of phosphates, three of oxalates, and one of cystin. Of the *compound* calculi, seventy-one were composed of uric acid or urates combined with phosphates; ten of uric acid or urates combined with oxalates; seven combined uric acid or urates, oxalates and phosphates; six of oxalates combined with phosphates. In one hundred and thirty out of the whole number the nucleus was of uric acid or urates, in nine of oxalates, in one of cystin; in four the nuclei were foreign bodies (two slate-pencils and two needles); in one calculus the nucleus had probably been of an organic nature, as it was found to be hollow.*

From their hardness and flinty nature, the oxalate of lime calculi are least fitted for lithotrity, whilst, from their softness and friability, the phosphatic ones are best adapted to this operation. The "ring" of the calculus will often indicate to the practised surgeon, when he strikes it with a sound, the probable nature

---

* *Apropos* of my remarks on the specimens of calculi now in the Liverpool Museum (1880), the following extract from a paper published in the *Medico-Chirurgical Transactions* of 1820 will be read with interest :—"Liverpool, population of the Borough 94,376. House (Infirmary) opened in 1749. Summary of a letter from Dr. L. J. Jardine. 'During the last 10 years 8 or 9 have been cut. Patients during 1817, in, 1314; out, 1623. Attached to the Infirmary is a Lunatic Asylum, which contains 86 persons. Dr. Dobson says : I have, from the observation and experience of 20 years, ascertained that in this place and the western parts of Lancashire, stone is an unfrequent disease. Out of 26,073 hospital patients, 6 only have been cut; that is, 1 in 4,345.'"—"A Statistical Enquiry into the frequency of Stone in the Bladder in Great Britain and Ireland," by Richard Smith, Esq., Senior Surgeon to the Bristol Infirmary. 1820.

of its composition; the sharp "click" of the hard stones contrasting with the duller "thud" of those composed of more or less phosphates.

When a stone is seized by the lithotrite, either for measuring or breaking it, the first grip or feel of it must not be regarded as deciding its composition throughout, as some of the hardest stones have very soft exteriors.

Rare forms of calculi, such as carbonate of lime and cystin, are occasionally met with. Of the latter I have given an illustration in what I believe to be the largest example known.

In determining the nature of the various calculi, you will find the table of Dr. J. Campbell Brown, Lecturer on Chemistry at the Liverpool School of Medicine, very convenient. I have his permission to append it.*

TABLE FOR THE EXAMINATION OF URINARY CALCULI.

1. Heat a portion of the powdered calculus upon platinum foil.

Destroyed. (a) *Uric acid : Ammonic urate : Cystine : Cholesterin : Bile-pigment.*

(b) Uric acid from *Calcic* and *Sodic urates.* Ammonia from *Triple phosphate.* Oxalic acid from *Calcic oxalate.*

Not destroyed. (c) *Calcic phosphate : Calcic carbonate.*

(d) Calcic carbonate from *Calcic oxalate* and *Urate.* Sodic carbonate from *Sodic urate.* Magnesic phosphate from *Triple phosphate.*

* *Liverpool Medical and Surgical Reports*, vol. iv., 1870, p. 43.

If it chars and gives odour of burnt feathers, add to another portion a drop of concentrated nitric acid and evaporate to dryness: pink colour; cool, and add ammonia: purple colour; *Uric acid* or *Urates*. If the odour is peculiarly disagreeable, resembling carbon disulphide, dissolve in ammonia, and allow the solution to evaporate spontaneously; microscopic six-sided plates indicate *Cystine*. Mix another portion with lime; ammonia may be evolved from the *Urate* or *Triple phosphate*.

2. Ignite another portion in the blowpipe flame until it burns entirely away—Class (*a*), see above—or leaves a white residue. If it fuses, it consists of the mixed *Phosphates of Calcium, Magnesium,* and *Ammonium.* Place a portion of the residue on red litmus paper and moisten with a drop of water; alkaline reaction indicates *Soda* or *Lime*, from Class (*d*) or from *Calcic carbonate*. Dissolve the rest of the residue, remaining after ignition, in water, and filter. If the filtrate is alkaline, add a drop of hydrochloric acid, and evaporate cautiously to dryness; microscopic cubical crystals prove the presence of *Sodium*. Dissolve any residue insoluble in water, with hydrochloric acid, observing whether or not effervescence due to carbonic acid takes place; add a comparatively large quantity of ammonic nitromolybdate, and heat; a yellow precipitate indicates *Phosphoric acid*.

3. Boil a portion of the powdered calculus in dilute hydrochloric acid; effervescence indicates calcic carbonate; filter; neutralize the solution by ammonia, and add acetic acid in excess; a turbidity remaining indicates *Calcic oxalate*. To the clear solution (or to the filtrate if calcic oxalate is present) add ammonic oxalate; a precipitate indicates *Calcium, which was not previously in the state of oxalate;* filter, if necessary; add ammonia, and stir; a white crystalline precipitate indicates *Magnesic phosphate*.

The size of calculi varies much, from the small

bodies which appear in the urine, and are spoken of as "gravel," up to masses weighing many ounces.

When we consider that all stones originally were small, too much importance cannot be attached to their early detection and treatment. The crushing of a stone that has recently made its escape from the kidney into the bladder is an operation as certain as it is safe; and after an attack of renal colic, which has probably been caused by the passage of a calculus, unless there is evidence of the stone having escaped naturally by the urethra, the bladder should be explored and the stone, if detected, crushed.

In this way I have on several occasions prevented a renal calculus from assuming larger proportions, and where its subsequent removal would have necessitated a correspondingly more serious operation. A well-marked instance was in a patient whom the late Mr. Long attended for an attack of renal colic. After the attack was over, Mr. Long felt sure that the calculus still remained in the bladder, and asked me to crush it. I found the calculus as it was predicted, completely broke it up, and removed the fragments with one of Bigelow's tubes. The patient was under the influence of chloroform, and so little was he inconvenienced by the operation that he was practically well as soon as it was over, and left for the country in three days. I was rather astonished at the size of the calculus, which was over half an inch in diameter; no wonder that its passage along the ureter was attended with the most acute suffering.

I feel sure that the ureter, though so small a tube,

is capable of giving passage to calculi of a far greater size than is generally supposed. Here is an illustration of a calculus, from a specimen in the Museum, which, though weighing 95 grains, almost succeeded in making its way into the bladder.

Those who are interested in cases—which in the present day, owing to the improvements in surgery, are rare—where enormous calculi have been removed by operation, will find numerous illustrations in the writings of surgeons in the earlier part of this and in the last century. Notably amongst them is one by Mr. Earle, which contains a record of some remarkably large stones.*

The smaller calculi not unfrequently indicate very distinctly that they are of renal origin, as casts of the uriniferous tubes can be traced upon them. Large calculi require some modification in the mode of effecting their removal, therefore it is of importance to determine their dimensions as accurately as possible. For the removal of very large stones the suprapubic operation has been successfully practised.

The measuring of calculi will be referred to in speaking of sounding for stone.

The shape assumed by calculi is sometimes very irregular and even fantastic. Here (Plate E, Fig. 1) is a remarkable specimen where the calculus is largely

---

* "Remarks on the Danger of Extracting Large Calculi," by Henry Earle.—*Medico-Chir. Trans.*, vol. xi.

An account of some very large calculi removed will be found in a quaint old work designated, *A Compleat Treatise of the Stone and Gravel, with an ample discourse on Lithontriptick, or Stone breaking Medicines.*" By John Greenfield, M.D., of the College of Physicians, London, 1710.

PLATE E.

made up of spike-like processes. It is from a patient who was admitted into the Infirmary in a dying state, and in whose bladder I discovered it, having reason to suspect that he was suffering from stone. The patient's condition was, in the opinion of my colleagues, in which I entirely concurred, so hopeless as to render any operation inadmissible; death occurred within twenty-four hours of his admission. It will be seen that one of the spikes is broken off; this I believe was occasioned by the sound. The calculus weighs three drachms. In reference to the shape assumed by calculi, I have two specimens to exhibit which in this respect are almost unique.

The first was removed by Dr. Lowndes, from a boy, at the Liverpool Northern Hospital, and weighs one ounce and half a drachm; the anterior portion of the calculus seems to have been cast within the prostatic portion of the urethra, the urine escaping by the side of it. (Plate E, Fig. 2.) The lateral operation was performed, and the patient made an excellent recovery.

The other specimen was successfully removed by Mr. Rushton Parker, at the Stanley Hospital, from a patient who had suffered from symptoms of stone extending over a considerable period of time. Here the calculus appears to have occupied the place of the prostate, only the capsule of the gland remaining; the urine not, as in the previous instance, escaping by the side, but through a channel in the stone, corresponding in direction with the continuance of the urethra.*

* *British Medical Journal*, January 19, 1878.

(Plate E, Fig. 3.) It weighs nine drachms and two scruples.

A case somewhat similar to Mr. Parker's is recorded by Mr. Sympson, of Lincoln, where a distinct channel was made in the calculus through which the urine passed.* Some very remarkably shaped calculi, cast, as it were, partly within the bladder and partly within the urethra, are described by the late Mr. Poland.†

Calculi occasionally present certain surface markings known as facets. These usually indicate the presence of other stones. In a case recorded by me, I removed, by lithotomy, from a child, two uric acid calculi, weighing respectively sixty and sixty-five grains. On examining them, they each presented three even facets corresponding to their three sides, each facet being about the size and shape of a fourpenny-piece. The explanation clearly was that the stones, lying side by side, had regularly rotated, remaining in contact sufficiently long to produce the markings described.

In a case where Dr. Rawdon removed two calculi from the bladder, one was marked with three facets and the other with two, the explanation being the same as that stated in my own case.

Stones are occasionally met with encased in a thick leathery fibrinous material, containing particles of phosphatic deposit. Illustrations are given of these in connection with the subject of sounding, where,

---

* *British Medical Journal*, March 23, 1878.

† *Guy's Hospital Reports*. Series iii., vol. iii.

from the nature of their composition, difficulty was experienced in their detection.

Lastly, I will notice concretions found upon bodies other than those on nuclei which may be derived from the urine or the blood. Examples of incrustations forming on foreign bodies introduced into the bladder, such as pieces of bougie, needles, etc., will be found in another place, where the management of accidents of this kind is treated of. I wish here more particularly to refer to the very important series of cases which are recorded in connection with gunshot wounds of the pelvis.

In alluding to some of the curiosities of stone formation, I may possibly be considered as digressing somewhat from the everyday practice of surgery, but though you may not be called upon to engage in military practice, yet none the less will you be liable to witness the consequences of similar casualties, the result of accident or of design.

The unfortunate conflict in America, which was attended with an almost unparalleled amount of destruction and mutilation to those engaged in it, most abundantly illustrates that of which civil practice is not entirely wanting in examples. In addition to the vast opportunities of observation afforded in that war, I am fortunate in being able to refer to clinical records which, not only in their extent, but in their accuracy and detail, are a credit both to the Government and the individuals that produced them.

Through the courtesy of my friend, Dr. Cushing, of Boston, I received a copy of these voluminous works, which I have had the pleasure of adding to the Library of the Liverpool Medical Institution.* I quote from the work referred to the following *resumé* of the recorded cases. "In the twenty-one lithotomy operations, seventeen were successful, three fatal, and in one the result has not transpired. Of thirteen cases where missiles were removed, there were ten in which these were leaden bullets, three of the round and seven of the conical variety; six of the ten balls were very slightly encrusted, while four formed the nuclei of large stones. In three cases the projectiles were of iron, a canister-shot, a grenade fragment, and an arrow-head, all coated with thick calcareous depositions. In eight cases in which bone, cloth, hair, or soft organic matters had constituted the nuclei, the calculi were of medium or large dimensions, and commonly very friable. In six cases of the last series, of what may be called traumatic calculi, there were no obvious contra-indications to lithotrity. In all of the encrustations and concretions the ammoniaco-magnesian phosphate prevailed, and several were almost exclusively composed of this triple salt; in others, phosphate of lime, urates, and organic matters were present in limited proportions. The remark of Marcet,† that vesical concretions of this sort are uniformly of the fusible species—composed, that is, of nearly equal proportions of phosphate of lime and of the triple phosphate of ammonia and magnesia—is not sus-

---

* *The Medical and Surgical History of the War of the Rebellion.* Washington, U. S. A.

† *On Calculous Disorders.* 1817.

tained by my observations, which rather tend to show that in such concretions the bone-phosphate is often altogether absent, and that the triple phosphate uniformly predominates." And what is true of missiles introduced into the bladder is equally true of other foreign bodies similarly placed. The records of both hospital and private practice show many illustrations of this — of calculi formed on necrosed bone, on bougies, catheters, and such-like articles capable of providing a nucleus. Of these further instances will be found in my lecture on foreign bodies introduced into the urethra and bladder.

TWENTIETH LECTURE.

Calculous Disorders—Stone in the Kidney—Renal Colic—Calculi impacted in the Ureter.

In considering the disorders of the urinary system, there is no subject of greater interest to the practical surgeon than the treatment of the various calculous concretions which are met with in all parts of the tract from the kidney downwards.

Calculous depositions, passing through the kidney under the name of " an attack of the gravel," most frequently come under the attention of the physician; small particles of sand-like material, giving rise to irritation and severe pain, are discharged with the urine, and relief follows.

There is no doubt that the nuclei of the majority of calculi are originally formed in the kidney and pass down into the bladder, where they are retained. There they increase in size, until, by producing signs of irritation, they give indications of their presence.

In illustration, I will mention a case where a uric acid calculus was passed, weighing ten grains, upon which could be seen casts of the calyces of the kidney. For some time the patient from whom it came had suffered from obscure renal symptoms, leading to the suspicion that a calculus was impacted in the kidney—a condition described as nephritic colic.

Suddenly these indications became greatly aggravated: there was intense lumbar pain extending down the groin and thigh of one side, with retraction of the testis. After a few hours of most excruciating suffering, these symptoms disappeared quite as abruptly as they commenced, and complete relief followed. There could be no doubt that a calculus had passed from the ureter to the bladder. A few days after, I sounded the patient, and detected in the bladder the calculus I now show you. I arranged to crush it with the lithotrite, but on the day following my examination it was voided with some pain and difficulty. Had I then done as I usually do now, sounded with my lithotrite, I should have spared the patient even this annoyance. If this stone had been allowed to remain in the bladder, it would have rapidly increased in size, eventually necessitating the performance of either lithotomy or lithotrity.

An attack of nephritic colic, abruptly terminating in relief, is a symptom sufficiently distinctive to require the examination of the bladder in search of the calculus, which most probably has descended. A few days may be allowed to see if the stone is naturally expelled. It is, however, under such circumstances, no kindness to the patient to encourage him, when the pain and suffering are over, to lay the flattering unction to his soul that his troubles are ended. If the calculus be not expelled from the bladder naturally, or be undetectable with the sound, the inference is that it has merely changed its position, and is still retained within the kidney or ureter, where it should be attacked with medicines and diet, with the view of exercising a solvent action upon it.

To relieve the urgency of the pain attending an attack of renal colic, due to the presence of a stone, opiates and warm emollient applications are indicated, in addition to the obvious mechanical expedients of giving copious diluent drinks and practising shampooing of the loins. The stigmata of maize is a remedy which has recently been introduced for the treatment of gravel and nephritic colic. By some observers it is supposed to exercise an anæsthetic action on the mucous membrane of the urinary tracts, whilst others lay stress on its service as a diuretic. The syrup appears to be the most reliable preparation.[*] Turpentine, in small doses, has also been recommended for a similar purpose.

There is a plan of favouring the expulsion of small calculi after their descent into the bladder which is by no means a bad one. It consists in making the patient lie on his belly, so as to bring the calculus on to the anterior wall of the bladder, then causing him to rise on "all-fours," and thus to micturate. By this means the stone, not falling into the depression behind the prostate, may be swept out by the stream of urine. I have known this manœuvre answer on more than one occasion. Napier's convolvulus catheter has also been ingeniously adapted for a similar purpose.

If the calculus remains in the kidney it is seldom

---

[*] *The Practitioner*, June, 1880, p. 452.

latent; it either grows and quietly hollows out the kidney structure, or it sets up pyelitis or abscess: if it becomes impacted in the ureter, it may lead to the complete atrophy of the corresponding kidney; and if it lodges in the bladder, it is simply biding its time until it shall have arrived at dimensions suitable either for lithotomy or lithotrity. The importance of attending to and following up the earliest symptoms of gravel or of stone cannot well be exaggerated.

A case similar to the one I have narrated, and showing the advantages of early exploration and treatment, came under my notice a short time ago. A gentleman, after suffering a paroxysm of pain, evidently due to the passage of a renal calculus, consulted me for persistent retraction of the testicle, with some tenderness. I introduced a lithotrite for him, and could distinctly feel, on entering the bladder, that I had pushed back a small calculus; this I seized and crushed. The patient passed some few portions of a uric acid calculus, and suffered no further inconvenience.

Calculi impacted in the kidney sometimes attain a very considerable size, and usually eventually give rise to inflammation, suppuration, and disorganisation of the affected organ. Such a case was under my care. The patient was sent to me by my friend, Dr. Fourness-Brice, then of the Cunard Service, who having diagnosed the existence of stone in the kidney, thought it might be possible to afford relief by operation. Unfortunately, however, the patient was in such broken health as to render any such course inadmissible.

Dr. G. G. S. Taylor, who made a post-mortem examination after the patient left the Infirmary, was good enough to inform me that the right kidney was found to be completely disorganised, and to contain a large quantity of calculi, varying in size from a pea to a hazel-nut. The left kidney was similarly affected, but to a less extent, thus confirming the view that had been taken as to the uselessness of resorting to nephrotomy.

Here is another specimen, which was also removed from a patient of mine in the Royal Infirmary, who died shortly after his admission, without giving me the chance of affording relief by an operation, which some time previously might, I believe, have been performed with a fair chance of success. The operation of nephrotomy is not one that presents any peculiar difficulty, as the kidney can be exposed in the way employed for the performance of colotomy. You will find an instructive paper on this subject by Mr. Thomas Smith.*

The diagnosis of stone in the kidney may be directly verified by the use of an aspirator needle. The first reported case of this kind is, I believe, by Mr. Barker.† I can add further testimony to the value and safety of this proceeding.

A calculus in the kidney or the commencement of the ureter may lead to suppurative nephritis and pyelitis, and the formation of an abscess, which may burst spontaneously and leave a urinary fistula. Such an abscess may require an incision in the loins to

---

* *Medico-Chirurgical Transactions*, 1869.

† *The Lancet*, April 17, 1880.

evacuate the matter. A condition similar to that which might be caused by the presence of a calculus within the kidney or ureter was observed in a case in which I saw Mr. Chauncy Puzey perform nephrotomy at the Northern Hospital, and to which reference is made in the lecture on the surgery of the kidney.

A calculus impacted in the ureter may lead to complete atrophy of the corresponding kidney; such cases, however, are by no means necessarily fatal, inasmuch as the opposite organ compensates for the damage by adapting itself to increased work. There is an interesting case recorded by Dr. Newman,[*] where death followed suppression of urine which had existed for five days. At a post-mortem examination, symmetrical blocking of both ureters with calculi was found. This, of course, is a very unusual instance.

I once saw a case with Dr. Davidson, at a consultation in which we had the advantage of Sir William Gull's presence, of complete suppression of urine occurring in a gentleman in whom there was the strongest evidence of a renal calculus in one kidney. The question as to whether both kidneys were similarly impacted was considered, but not entertained, the history pointing to the conclusion that it was the secreting action of the kidneys which was in suspense. This conclusion proved to be correct, and the patient eventually made a good recovery. Though there could be no doubt that one kidney contained a calculus, its presence did not, I believe, determine the suppression, which nearly proved fatal. In connection with

[*] *British Medical Journal*, Jan. 15, 1875.

the subject of renal calculus, this case has a special interest.

Through the kindness of Dr. Rawdon, I had the opportunity of seeing with him and examining a patient upon whom he had performed lithotomy and in whom the presence of a calculus within the ureter was diagnosed in a manner that I had not previously noted—namely, by the finger in the rectum. The following are the notes of this case:*—

T. F., aged six, was admitted into the Liverpool Infirmary for Children on September 20th, 1878, suffering from symptoms of stone in the bladder, extending over upwards of two years, during which period he had on several occasions passed pieces of gravel. On introducing a sound into the bladder, a calculous body could be felt, which gave a grating sensation rather than the characteristic "click" of a calculus as felt in children. The urine contained a good deal of mucus, and at times a few gritty particles.

On September 28th, lateral lithotomy was performed, and three portions of broken-down calcareous deposit of a phosphatic nature were removed by the scoop, and the bladder then cleared.

For the first week, the wound assumed an unhealthy appearance; the urine was scanty and offensive, and at times contained calcareous deposit.

After the ninth day, both the wound and general condition of the patient improved, and convalescence appeared to be steadily advancing. On the twentieth day, no urine was passed for six hours. This condition was remedied by a diuretic and demulcent drink.

On the morning of the twenty-seventh day he was taken alarmingly ill, was collapsed, and complained of pain low down on the left side. In the afternoon he had a sharp rigor. Under

* *British Medical Journal*, Feb. 1. 1879.

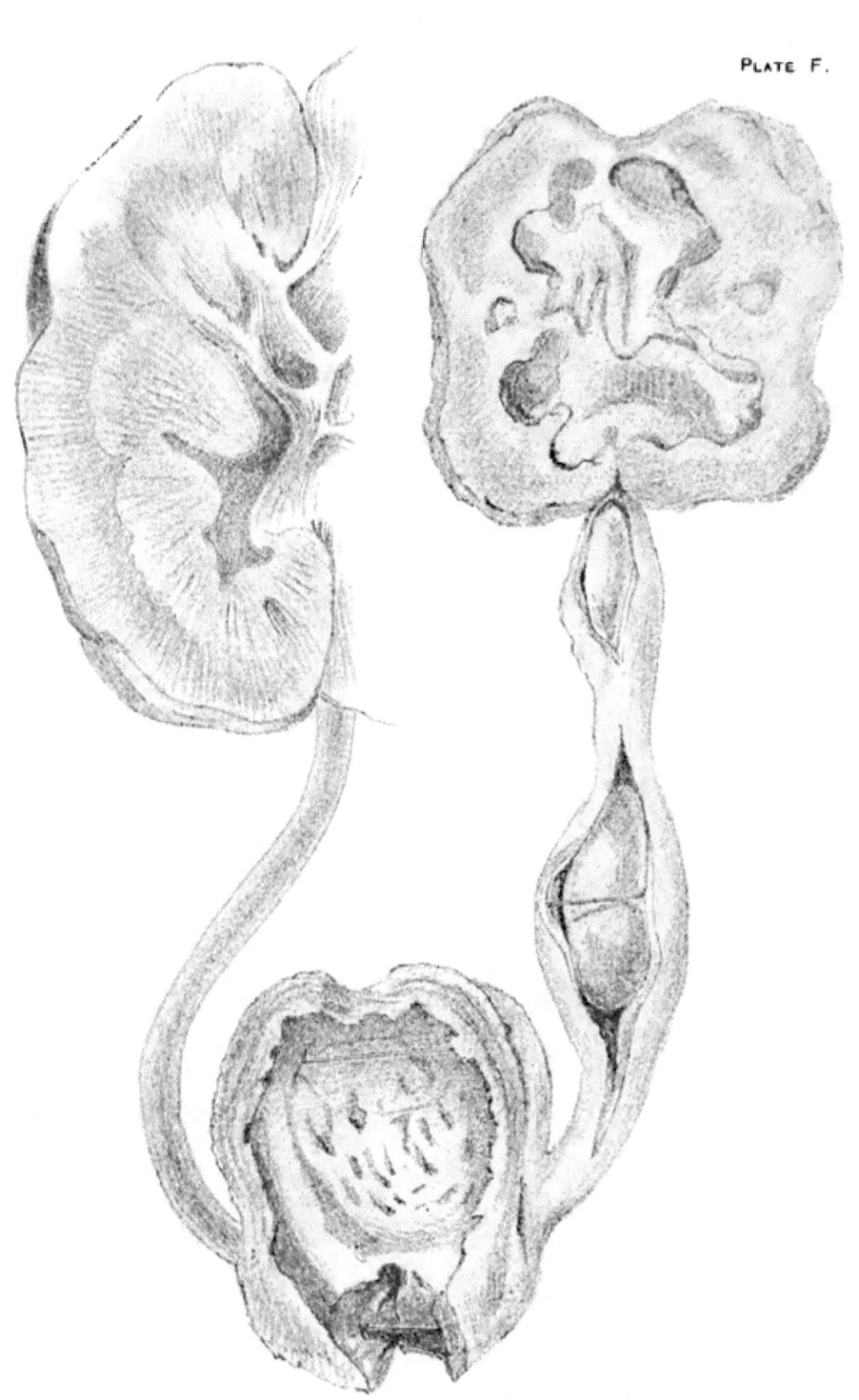

PLATE F.

chloroform the bladder was explored, but nothing was discovered here to account for these sudden symptoms. The rectum was then examined by the finger, and a solid body was felt, apparently fixed to the bladder, on the left side of its posterior wall. This body was diagnosed to be a calculus in the left ureter.

From the twenty-seventh day, although the patient had no second rigor, he suffered occasionally from vomiting and diarrhœa. His temperature rose to 103° in the axilla, his pulse became rapid (170) and feeble, and he had well-marked tenderness upon pressure over the left loin.

He sank (October 30th) on the thirty-second day after lithotomy.

The right kidney was hypertrophied, being at least a third over the normal size; in structure it was healthy. The ureter was natural. The left kidney was atrophied, being one-third under the normal size. The secreting structure was the subject of chronic degenerative change. The pelvis and the calyces were considerably dilated. A diffuse perinephritic abscess was present, containing from two to three ounces of thin, fœtid, urinous pus. The abscess extended downwards along the ureter. The ureter (the left) was distended, throughout its length from kidney to bladder, to the size of a piece of small intestine, being impacted with friable cretaceous substance, in which two true calculi were imbedded. The larger, which resembled a date-stone, was close behind the orifice of the ureter. The walls of the bladder were thickened, but otherwise healthy. There was no peritonitis.

I have given a sketch (Plate F) of the pathological appearances observed in this case.

TWENTY-FIRST LECTURE.

SYMPTOMS OF STONE IN THE BLADDER — SOUNDING — THE MICROPHONE — SOURCES OF ERROR IN SOUNDING.

I WILL now pass on to notice the symptoms of stone in the bladder. Collectively they may be described as those indicating vesical irritation, but as there are other causes for this symptom it will be necessary for me to refer to them individually.

A state of irritability of the bladder is one of the commonest consequences of the presence of a stone within it. In children a calculus may be the cause of nocturnal incontinence, the patient appearing to lose all control over the act of making water.

Pain is met with under different circumstances. There is usually some dull aching felt about the neck of the bladder, increased after micturition. Pain is also not unfrequently referred to more distant parts; it is common to hear patients describing it as being at the end of the penis, to relieve which I have seen the prepuce greatly elongated and hypertrophied by manipulation. When you hear of children showing signs of an irritable bladder, and complaining of pain at the end of the penis, you may be sure they are suffering from stone. An adult may suffer in the same way, the symptoms giving rise to the suspicion of stone, but being really dependent on a tender prostate.

Pain is also generally increased by sudden jolts or actions of the body, and in some recorded cases such accidents have been followed by the unexpected appearance of stone symptoms; the explanation being that previous to this the calculus had occupied some part of the bladder where it had not caused any inconvenience until its position had thus been altered.

Hæmaturia is a frequent symptom of stone, the blood flowing, often nearly pure, just as micturition is finishing. It was the only symptom of much moment in one case that was under my notice where I recently performed lithotrity. A gentleman alighted from a railway carriage before the train had stopped; the concussion that this occasioned was followed shortly afterwards by the passing of about half an ounce of blood, and led to the suspicion of stone, which subsequent examination verified.

Prolapse of the rectum is a common symptom in children; it is caused by straining to empty the bladder. Where it exists we should not forget to enquire into the state of the urinary organs, unless it can otherwise be explained.

The condition of the urine may also indicate the probable presence of some source of irritation within the bladder.

The existence of the symptoms of vesical irritability to which I have referred, either collectively or individually, will warrant us in proceeding to obtain that conclusive evidence which sounding alone can afford.

Before proceeding further, I would remark upon the

fact that occasionally we meet with cases in which calculi, and not small ones either, appear to have given rise to no symptoms whatever, their discovery being brought about by some such accident as that I have mentioned in connection with sudden movements causing pain or hæmorrhage. There are some people we meet with who are particularly insensitive to morbid impressions, and stones and strictures may in them go on forming until some casual circumstance arises to quicken their observation in respect to themselves. It is sufficient for me to mention here the disproportion that seems to exist in a few individuals between disease and its symptoms.

Conclusive evidence of the presence of a stone in the bladder can only be afforded by the employment of those means of exploration to which the term "sounding" is applied.

The process of sounding is a far more delicate and deliberate proceeding than at first sight we may be disposed to admit.

Formerly, when there was only one method of removing a stone, the determination of its presence was the simple preliminary, and its removal by lithotomy naturally followed. In the present day, with lithotomy, lithotrity, and litholapaxy—and, perhaps, soon we shall have to add, lithontriptics—we are bound to obtain precise information also as to the dimensions and physical constitution of the foreign body before we are in a position to consider how its removal will be best and most safely effected.

A careful and accurate sounding not only determines what is proper to be done, but by anticipation obviates many of the difficulties which may arise in carrying out the appropriate method of treatment. I have frequently had in practice illustrations of this.

Being satisfied that the symptoms of vesical irritation which a case may present cannot be traced to their source until the bladder has been directly explored, it will be necessary to proceed to obtain this information in the fullest and most precise manner.

And in sounding I would urge the same caution as I have already pointed out as being necessary in all instrumental examinations of the urethra, such as the passage of catheters and bougies, where a short delay is a matter of no moment. I have known more than one instance where sounding has been followed by urinary pyrexia and death, post-mortem examination revealing not only what the sounding had ascertained—*i.e.*, a stone in the bladder, but what it had not—viz., disorganized and suppurating kidneys. I do not mean to say that the sounding might have been dispensed with, or the pyrexia warded off, but a more complete examination of the health of the whole urinary apparatus would have indicated at once the degree of risk that was incurred, and the necessity for additional precaution in preventing any ill consequences. When a case is admitted into my wards in the Infirmary, my directions are, First examine the urine, and then I will sound. When there has been any degree of kidney complication, rest in bed, warmth, diluent drinks, and doses of aconite (Fleming's tincture) or quinine, have, I am sure, frequently prevented, or moderated, unto-

s

ward symptoms. To sound persons in your consulting room, probably seeing them for the first time, and perhaps fatigued by a journey, is, to say the least, incurring a risk which is generally unnecessary, and which you may have cause to regret.

In sounding a patient it is best to employ an anæsthetic; I generally use ether, as I do for most other operations. It is quite impossible for a surgeon to obtain all the information about a calculus that is requisite with the patient writhing or resisting; whilst his insensibility greatly facilitates the object you have in view—namely, precise, not general, information. In the only case of stone that was examined in my ward without an anæsthetic, I had cause to regret the omission, as the boy was so frightened that he ran away, and could never be found, though I had him sought for most carefully. I thus lost the opportunity of freeing him from a disease which ever afterwards he would endeavour to conceal, until his symptoms became unbearable, if not fatal.

In proceeding to sound, let it be remembered that the instrument we use is merely an imperfect substitute for the finger, and to obtain the information required it will be necessary to conduct the operation with precisely the same method as we should adopt in the digital examination of any cavity or space in the body which can be so reached. To pass a sound into the bladder, and then indiscriminately wriggle it about with the hope that it may strike a stone or reveal a rugged or ulcerous surface, is a proceeding as hazardous as it is useless. This latter mode of manipulating you will not employ if you remember that your sound or lithotrite is only a very imperfect substitute for your finger.

First of all, in reference to the selection of a sound, I generally use a small lithotrite, with which I can both feel and measure.

Of sounds proper, I prefer one with a short curve and a somewhat expanded extremity. Nearly all sounds have one grave defect—they are made too thick in the stem, so that they almost *fit* the urethra. To sound a bladder without producing spasm or hæmorrhage, you require an instrument which lies *loosely* in the urethra, and can be moved freely in every direction. A thick-stemmed sound gives you no better information than a thinner one; the sensation produced to your ears and fingers on striking a stone in the bladder being precisely the same in both cases. With such a sound as I have described, and a normal bladder, there ought to be no difficulty in finding a stone.

There are, however, contrivances which have been suggested having for their object the increasing of the sensitiveness of the instrument. Of these, I may mention Napier's sound, where the bulb is made of a softer metal, capable of receiving scratches, such as contact with a calculus is likely to make. The advantage gained by the chance of taking impressions by the sound is, however, probably counterbalanced by its failure to elicit a clear note, such as a steel instrument is capable of producing.

Dr. Duncan has, I think, improved upon this idea,

by coating the end of a steel sound with some carbonaceous material, as soot, which is then varnished and dried. Such a sound is capable of giving both ocular and audible evidence of its contact with a calculus.

Lastly, we have the microphone. Shortly after Sir Henry Thompson published his remarks in reference to the application of this instrument as a sound for stone,* I was engaged, in conjunction with a practical electrician, in testing its value for the purpose suggested. The conclusion we arrived at, from a series of trials, was, that the microphone in its present form was not likely to give us any information we could not obtain by the employment of the other means to which I have alluded; I therefore think it unnecessary to repeat to you the experiments upon which this deduction is based. I introduce this reference, as I have been asked for information on this point.

The patient being placed under ether in the horizontal position, is now ready for sounding. One or two hard pillows should be at hand, in order that the surgeon may be able to alter the axis of the pelvis, should it be necessary, with the view of causing the bulbous extremity of the sound to pass over the whole surface of the bladder. Where there is an enlarged prostate, especial care should be taken, by rotating the handle, to make the bulbous end completely sweep the dip behind the enlarged gland, otherwise a calculus, thus concealed,

* *British Medical Journal*, June 8, 1878.

may avoid detection. Do not wriggle the instrument about, but go over, methodically, every part of the bladder in the course of your search. If there is no stone, there may still be a morbid condition of the mucous membrane sufficient to explain the symptoms complained of.

The necessity for sounding the bladder in a complete and systematic manner is shown by cases occasionally being met with where the stone occupies some unusual position. A calculus retained in the orifice of the ureter is capable of producing symptoms precisely similar to those significant of its presence in the bladder. Such an instance once presented itself to my notice, where I believe a stone was impacted in the vesical opening of the ureter. At all events, sounding served to dislodge a calculus which was distinctly felt in a position corresponding with the opening of the ureter. It was passed spontaneously two days after my examination, and complete cessation of the symptoms complained of ensued. Guthrie records a case in which, after death, a stone was found sticking in the orifice of the left ureter.*

Should a stone be found, its presence will probably be recognised both by the touch and the ear. A stethoscope applied over the pubes renders the sound more audible, and may be resorted to if there be any doubt.

A stone having been discovered, we proceed to its further examination, with reference more especially to its size and formation. As I have previously said,

* *Op. cit.*, p. 6.

where I can, I sound with a lithotrite, so that I may at once take the measurement of the stone, and, by seizing it, judge of its composition. If, as in a small child, the lithotrite is not applicable, I endeavour to obtain this information by manipulating the stone with the sound. By a little practice, combined with method, you can in this way be very accurate in your conclusions. By the same means you must try to ascertain whether more than one stone is present; this is comparatively easy when a lithotrite can be used, but is difficult and uncertain with the sound alone.

The probable nature of the calculus is indicated partly by the character of the note that is obtained; the dull "thud" of the phosphates is as characteristic as the sharper "click" of the oxalates or urates; and, further, the examination of the urine often adds evidence which is conclusive.

Where there is a suspicion of the existence of a stone, which yet cannot be detected by sounding such as I have described, it will be well to make an examination on a subsequent occasion, with the patient in a different position. The sound may be passed with the patient standing and leaning forward, resting his hands on the back of a chair, with his legs apart. Again, I have sounded a man lying on his belly instead of his back.

The difficulties which arise in making a diagnosis of stone in the bladder are for the most part traceable to the existence, as a complication, of one of the three following conditions:—

First: the presence of a stricture of the urethra, or of an enlarged prostate.

Second: a diverticulum, or recess within the bladder, in which a calculus may be lodged.

Third: the coating of the stone with an imperfectly organised leather-like substance, which conceals it from detection with the sound.

In all cases of long-standing stricture it is well to bear in mind that the symptoms of stone may be concealed by those of the stricture; this occurred in a case of Mr. Stubbs', to which I have already alluded, where a stone was not detected or even suspected until perinæal section had been performed. Under similar circumstances, I crushed with the lithotrite, some time ago, a small calculus, which I only discovered just as I was completing the treatment of a stricture by Holt's operation. I have already drawn attention to the fact, an illustration of which is given in (Plate D), that an enlarged prostate may serve to conceal a calculus, unless special care is taken in using the sound to avoid such a source of error.

As an example of the difficulties arising from the presence of diverticula, or recesses within the bladder, in which a stone may be concealed, I will refer to an interesting case narrated by Mr. Hakes, at the Liverpool Medical Institution, during the session of 1863-64.

The patient was an old man, suffering from stone in the bladder, for which lithotrity was performed at the Northern Hospital; unfortunately, the operation terminated fatally. On making a post-mortem examination, a diverticulum, larger than the bladder itself, was found, communicating with the floor of the bladder by an opening admitting the little finger. In it was contained a portion of the calculus, broken by the lithotrite. Such a condition as this might not only account for a calculus being undetected, but would cause serious difficulties in the performance of lithotrity. Digital examination by the rectum, after the bladder had been emptied with a catheter, would probably afford the best means of detecting this complication, should any suspicion of its existence arise.

As bearing upon this difficulty in connection with sounding, I will quote the following remark :—" It will naturally be asked, How can this state of sacculation of the bladder, with or without stone, be recognized during life ? It must, I fear, be admitted that the indications are few and unreliable. By noting that a man, having an enlarged prostate, makes water slowly and with considerable exertion; that when a catheter is used, and after the bladder is apparently emptied, there is still a further flow; and particularly if the appearance of the urine during the double flow vary considerably, we may infer that cysts or sacculi do exist, although we cannot surely know it."*

Examination of the supra-pubic region with the hand should also not be omitted. I was reminded of the importance of this on looking through the specimens in the Museum of the City of New York Hospital. Appended to one (784), where there was a sac larger than a hen's egg opening into the bladder near the

---

* "Sacculation and Stone in the Bladder." W. Cadge, *British Medical Journal*, October 2, 1875.

fundus, in which were several calculi, is the note, "These calculi could not be detected by the sound during life, but the pouch containing them could be felt through the abdominal parietes."

Digital examination of the bladder by the hand in the rectum will be referred to in connection with the subject of tumours of this organ.

In the last place, the stone may be so constituted as in itself to oppose a difficulty in detecting it by the means usually employed.

In 1863, a boy was admitted into the Liverpool Royal Infirmary, under the care of the late Mr. Long, suffering from prolapsus ani, purulent urine, and painful and frequent micturition.

The child was sounded carefully, but without any evidence of stone being afforded. Death occurred in the course of a few days. On making a post-mortem examination, the kidneys were found in an advanced stage of disorganization. One was extensively sacculated, with its cortical structure nearly gone; the other was much enlarged and structurally changed. The bladder was small, and in it lay a calculus, made up of a urate of ammonia nucleus the size of a damson-stone, surrounded by a thick layer of soft material consisting of mucus, fibrin, and a little gritty matter, probably phosphatic. The outer covering could be cut or torn easily; and, after it had been in spirit, it presented, on section, a laminated appearance, like the fibrinous layers found in an aneurism. On striking the mass with a metal instrument, no ring was produced; hence the impossibility of determining its existence with the sound during life. I have seen something similar, but less well marked, in other cases where lithotomy has been performed. The specimen just described was exhibited before the Liverpool Medical Society by Dr. Rawdon, and is now in the Museum of the School of Medicine. In the accompanying sketch the nucleus of the stone, with a section of the fibrinous laminæ, are seen. (Plate G.)

Mr. Bickersteth records a similar instance. "It was that of a boy, who had every symptom of a stone, but in whom repeated examinations gave no clear indication of its presence. When the sound was introduced, I could feel, with my finger in the rectum, some apparent thickening in the posterior part of the bladder. I operated, and extracted a mass precisely similar to that just mentioned (Mr. Long's case), and the child recovered."[*]

In the absence of what is regarded as the only positive sign of the existence of stone in the bladder— namely, its detection by the sound—reliance must be placed on other, though less worthy, evidences; of these, in children (amongst whom, so far as I know, only soft calculi have been found), the presence of prolapse of the rectum, together with signs of urinary irritation, would, in the absence of any other explanation, justify what might be regarded as an exploratory operation. The slight risk attached to the operation of lithotomy in children it would be justifiable to incur where there were reasonable grounds for suspecting that the

[*] "Observations on Lithotomy."— *Liverpool Medical and Surgical Reports*, vol. i., 1867.

PLATE G.

A Bladder
B Fibrous lamina
C Stone Nucleus

symptoms might be due to the presence within the bladder of such a calculus as I have described.

These, then, are illustrations of some of the difficulties which we may have to encounter in determining the existence of a stone in the bladder.

Calculi, or portions of them, occasionally become impacted in the urethra, and in children are the most frequent cause of retention and extravasation of urine. (This latter condition is referred to more fully in another lecture.)

There are one or two practical points in reference to the treatment of calculi impacted in the urethra to which I will allude.

In the first place, it is most desirable to avoid opening the penile portion of the urethra, except the meatus, for the purpose of extracting a calculus, should it be so fixed. A wound of the urethra here is often most difficult to heal, and frequently results in a permanent fistulous opening. If the stone cannot be extracted by forceps or other suitable instrument, it is better to push it back towards the bladder, and with the finger in the rectum to command it, whilst an incision in the perinæum is made through which it may be withdrawn. Such a wound heals kindly, as in median lithotomy.

Where a calculus is retained within the meatus—the narrowest portion of the urethra—should there be any difficulty in extracting it, it is advisable to incise the meatus rather than to lacerate it by the exercise of force; the former proceeding is less likely to be followed by a stricture.

When a stone has been removed from the urethra, it is very necessary to explore the bladder with the sound, as others may remain behind and form the nuclei of larger ones. In a case where I performed lateral lithotomy in a boy, at the Royal Infirmary, my attention was first called to him by having to remove from his urethra a small calculus, which proved on examination to be only a portion of a larger one. On introducing a sound, I detected a stone within the bladder, on the removal of which I found that a small spicula of it had become detached and found its way into the urethra.

TWENTY-SECOND LECTURE.

TREATMENT OF CALCULOUS DISORDERS—LITHOTOMY.

THE treatment of stone in the bladder resolves itself into deciding in a given case whether lithotomy or lithotrity shall be resorted to.

There is another method of treatment—by lithontriptics or solvents—which, though it has been the subject of much painstaking enquiry and experiment, particularly by Dr. W. Roberts, has not yet been brought into such a form as to render it practically useful, so far as urinary calculi are concerned, except to a comparatively limited extent, and this chiefly in the treatment of renal concretions, or in combination with lithotrity.

I think it will be generally admitted that nothing better illustrates the progress which the science and art of surgery have made within the present century, and in our own recollection, than the treatment of stone in the bladder; nor would it be profitless or uninteresting, if occasion offered, to trace the successive stages by which this advance has been made. Further than this, lithotomy and lithotrity, by their limitation to the class of cases to which they are respectively appropriate, have undoubtedly effected a considerable saving in human life.

The operation of lithotomy, from whatever standpoint we look at it, must be regarded as among the most successful of the proceedings of any magnitude we are called upon to execute; in illustration of this, I would mention that in the obituary notice of the late Mr. Southam,* it is stated " he had performed the operation of lithotomy one hundred and twenty times, and had only lost one patient."

What are the conditions which determine the selection of lithotomy or lithotrity? † In endeavouring to answer this question, I shall avail myself of the large mass of practical information which at the present day surgery affords, and draw from it such deductions as my own personal experience and observation warrant. To attempt to provide a solution applicable to every case, complicated or otherwise, would be hopeless; I must perforce confine myself to general principles which may serve to guide us.

And first of all, there can be no doubt that lithotomy is *the* operation for children, and until the development and growth of the urinary organs are completed and maturity is reached. The deaths, very few in number, following it in young subjects are, with hardly an exception, traceable to some concurrent disease, such as tubercle, or to an accident occurring in the course of the operation. Nor is lithotrity free from the latter

---

* *The Lancet*, May 5, 1876.

† "Each operation has its special province, the boundaries of which (if indeed they admit of being fixed at all) can be determined only by a comparison of a vast collection of facts, carefully noted, and, above all, faithfully reported and properly authenticated."—Sir Philip Crampton.—*The Dublin Quarterly Journal*, vol. i., 1846.

objection. Lithotomy is to be preferred where the stone is large. I define a large stone as one having a diameter of two inches or more. I have crushed a stone of this size, but it was entirely phosphatic, and readily amenable to the lithotrite. Lithotomy is to be recommended where there is good reason to believe that, with a large prostate, there is sacculation of the bladder.

Where, in addition to an hypertrophied prostate, we have an advanced state of cystitis, or an atrophied or a paralysed bladder, though these conditions add to the risk attending all operations on the urinary organs, they must not be regarded as precluding lithotrity.

Advanced kidney disease is alike unfavourable to lithotomy and lithotrity. If the stone were of a size that could be crushed and removed at one, not too prolonged, sitting, I should be inclined to advise lithotrity; if, on the other hand, it were of dimensions or composition not allowing of this, I should prefer the risk of lithotomy rather than that of pyelitis, which would be sure to follow the unavoidable retention of rough and irritating fragments within the bladder.

When the prostate is enlarged, the facility with which the stone can be caught by the lithotrite, when measuring it, determines in a manner the selection of lithotomy or lithotrity. If there is difficulty in seizing it, as a rule, lithotomy will be preferable, whilst if it is readily caught, and does not exceed two inches in diameter, lithotrity may be practised.

The only special preparation required by a patient who is to undergo lithotomy, is that which is necessary to secure an empty and collapsed state of the rectum. I prefer giving the patient an aperient a couple of days prior to, and an enema on, the morning of the operation.

There are various modes of performing lithotomy; I shall confine myself to that which is practised with so much success at this Infirmary—namely, Cheselden's, or the lateral operation.

Ether having been administered and the patient secured, either by bandages or Pritchard's anklets, in the lithotomy position, the staff is then passed. If previous examination with the sound has revealed the presence of the least hitch or difficulty in passing an instrument along the urethra, it is better to introduce the staff before the patient is tied up; that is to say, before his position is materially altered from that in which catheterism is usually practised.

The operator is now prepared to make his first incision, having entrusted the staff to the care of an assistant, with the instruction, should the latter not be accustomed to this duty, under no circumstances to withdraw it until he is distinctly told to do so by the operator. Neglect of this precaution—a misunderstanding between the operator and the staff-holder—has more than once, I am told, led to a patient being removed from the operating room without the extraction of the stone.

The surgeon should then pass his finger into the rectum and finally adjust the position of the staff, taking care that it is not so much depressed as to unnecessarily endanger the gut when the knife is used.

Remember that to reach the groove you have to cut, and no artifices, such as endeavouring to make the staff prominent towards the perinæum, are likely to be of any material assistance; on the contrary, by altering the natural relation of the parts they rather embarrass or misdirect the operator.

The close relation of the rectum, when distended by fæces, with the upper portion of the urethra is well shown in Braune's *Topographical Anatomy*, where the plates are drawn not when the parts are in a flaccid condition, but from frozen sections specially prepared for the purpose. The author remarks, "one is astonished at the narrowness of the space between the upper portion of the urethra and the rectum."

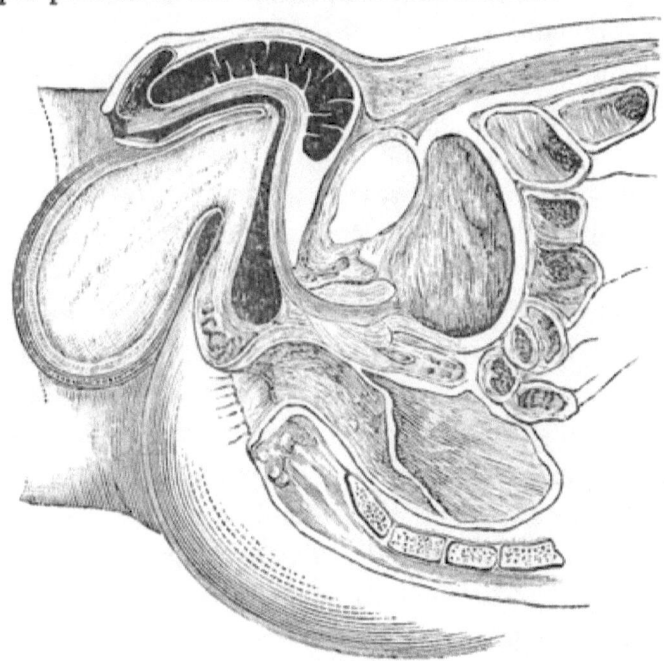

Fig. 34.

With the rectum empty, as it should be before lithotomy, the introduction of the finger favours its contraction, and thus tends to secure its safety.

Being satisfied that there is a stone in the bladder, that the end of the staff is also there, and that it has made its way along the legitimate passage, the operator is prepared to make his first incision.

Practically, to reach the bladder, two incisions are necessary. Though it is not always possible or even desirable to execute these by two movements of the knife only, yet, with a little practice, it is easy to do so. The object of the first incision is to open the perinæum and to reach the staff; that of the second, guided by the groove in the staff, to effect an entrance into the bladder.

Some difference exists amongst surgeons as to the precise point on the perinæum where the first or external incision should begin. The late Sir William Fergusson directed that the incision should commence one inch and three-quarters in front of the anus, a little to the left of the raphé, whereas Mr. Coulson and Dr. Keith say one inch before the anus, a direction which appears to be similar to that practised by Cheselden. In my earlier cases of lithotomy I adopted Fergusson's incision; I have, however, in my subsequent experiences gradually approached nearer the anus, and now concur with the directions of Coulson in considerably lessening the distance between the commencement of the incision and the anal aperture. For the gradual change in my mode of operating, I cannot assign any other reason than because by the latter method I find it easier and quite as safe to enter the staff, safer in fact, so far as the bulb is concerned. The commencement of my incision now almost corresponds in plane with the spot where I am desirous of striking the staff. On no other ground than that of natural selection can I explain the practice which I have gradually arrived at.

Commencing, then, an inch in front of the anus and a little to the patient's left side of the raphé, I deliberately plunge the knife, with a very slight upward direction, towards the staff as it lies in the membranous portion of the urethra. I usually touch the staff at this point, and, withdrawing my knife, complete the external incision and divide the perinæum obliquely, in a direction downwards and outwards, to the extent of about two and a half inches, the depth of the incision diminishing until, at its termination, the skin and superficial fascia are alone included. The incision should pass downwards about midway between the anus and tuber ischii. It is of some importance that the first incision should be sufficiently free; much of the difficulty I have occasionally seen in extracting the calculus or in arresting hæmorrhage has arisen from the operator making so limited an incision as to cause himself embarrassment in executing the necessary manipulations. Where the incision has been free, even where it has erred somewhat in this direction, I have never seen, or myself experienced, difficulty in tying any artery that may have been divided, including on one or two occasions that of the bulb. The left forefinger, which hitherto has assisted in steadying the perinæum, is now passed into the wound. A few fibres may require touching with the knife before the finger-nail can be passed into the groove of the staff. When the operator is certain that the point of his blade is well within the groove of the staff, the knife is steadily pushed onwards to the bladder, its edge being inclined more obliquely outwards than in the first incision, with the view of incising the prostate in a direction corresponding with its broadest axis. In this way the second, or deeper, incision is completed.

And now comes an important distinction to remember, in operating upon the child or the adult—a difference which requires a corresponding adaptation in the pushing of the knife along the staff into the bladder. There is a variation in the plane of the deeper portion of the urethra in the two periods of life referred to. In the child the bladder is an abdominal rather than a pelvic organ, and corresponding with this there is an upward obliquity of the urethra towards it which is not to be forgotten; care must therefore be taken when operating on children to depress the handle of the knife as the point is pushed onwards to the bladder. This is well illustrated by Mr. Bickersteth, in his "Observations on Lithotomy," to which reference has already been made.

The incision into the bladder having been completed, the left index finger should be insinuated along the staff in the opening that has been made. To permit of the tip of the finger passing into the bladder, dilatation may be employed; but no force, either in introducing the finger or withdrawing the stone, is to be exercised, otherwise the prostate may be torn, or the connections of the bladder loosened. The staff must not be withdrawn until the index finger is well within the bladder, when most probably the stone will be felt. "If possible, and it almost uniformly is so, the stone should be felt with the finger before any instrument is introduced or attempt made to seize it."*

In the case of adults with an unusually deep perinæum, I know it is impossible, even with a long index finger, to reach the bladder. In these cases an old-fashioned gorget will be found exceedingly useful in completing the incision along the staff into the bladder, and acting as a guide for the introduction of the forceps. Where such a condition of the parts exists, some surgeons pass the forceps by the side of the staff, after the deep incision has been made with the knife or the gorget, and do not withdraw the staff until the forceps touch the stone. Under no circumstances should the staff be removed if there is the slightest doubt as to whether the bladder has been fairly reached, as in difficulty it is our only guide. Directed by the left index finger, the forceps should then be passed into the bladder and the stone seized and extracted, without the exercise of anything like jerking or force. Remember, if in doubt, it is better to cut than to tear or to contuse indiscriminately, with the additional chance of loosening the bladder from its connections. Hence, in the extraction of large calculi, the prostatic incision may advantageously be extended

* *Practical Surgery*, by Robert Liston, p. 413.

when necessary. This may be done with a straight probe-ended knife, which should be at hand. In a case where I removed a calculus weighing over two ounces, I made a bilateral section of the gland. In a case of Mr. Hamilton's, at the Royal Southern Hospital, where he successfully removed by lateral lithotomy a calculus weighing nearly five and a half ounces, it is stated that the prostate was further incised in an upward direction.*

A scoop should always be in readiness at this stage of the operation, as the calculus may break, when it will be found useful in removing fragments. Some operators, amongst them Professor Humphry, of Cambridge, seem to prefer a scoop, assisted by the finger, instead of the forceps, for extracting the stone.†

Should the calculus break up, as phosphatic ones will sometimes do, it is well to wash out the bladder by means of a pipe introduced through the wound, and a Higginson's syringe, so that nothing may be left behind.

Calculous incrustations on the walls of the bladder need not be scraped off, as, when the urine returns, as it usually does after the operation, to its normal acid reaction, these deposits become washed away or disappear. After the stone has been removed, the finger should be re-introduced into the bladder, for the purpose of determining whether other calculi remain. The wound should then be carefully examined, and any accessible bleeding points tied. When the hæmor-

---

* *British Medical Journal*, Dec. 9, 1865.

† *The Lancet*, June 1, 1872.

rhage is of such a character that it cannot be stopped by ligature, the dilatable tampon of Mr. G. Buckston Browne will be found very efficient. It is a great improvement upon the old umbrella lithotomy tube, and can be most readily applied and removed.*

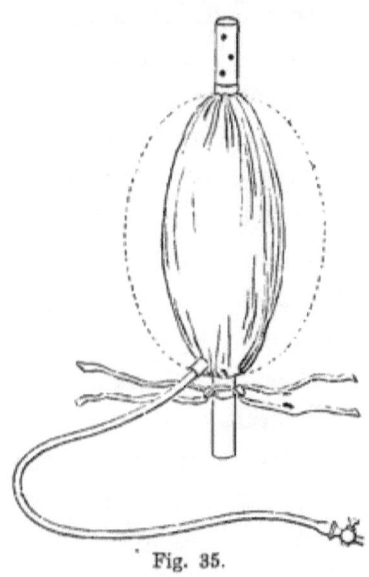

Fig. 35.

Formerly, I adopted the practice of introducing a gum-elastic tube through the wound into the bladder, retaining it there for twenty-four hours; more recently I have discarded it. I do not see there is any advantage in this provision, other than to prevent the blocking up of the wound with clots, an accident which is easily detected and remedied by those in charge of the case.

In the after-treatment of lithotomy, the late Mr. Southam, whose success in this operation was remark-

* *The Lancet*, Sept. 15, 1877.

able, attached great importance to preventing the injurious effects of shock. Immediately the operation was over he directed that the patient should be enveloped in a warm blanket, and hot bottles applied to the feet and other parts of the body, until reaction had set in.* In other respects the after-treatment of lithotomy differs but little from that of other wounds. As in all operations on the urinary organs, special care is necessary to keep the patient sweet and clean, and prevent the air of his apartment becoming contaminated; to mop up the urine as it is discharged from the wound, I have employed sponges wrung out in carbolic lotion, and constantly changed.

Urine usually begins to flow through the urethra about the tenth day, the first discharge *per viam naturalem* not unfrequently being followed by a chill or rigor, which, so occurring, need not occasion alarm. I allude to this, as it is about the only rigor following an important operation which is not to be regarded with a grave suspicion that it may forebode something worse.

As already stated, I confine my description to that method of operating which for over fifteen years I have successfully employed. So far, it has been found applicable to all periods of life, and to both small and large calculi. I can add testimony contradictory to a statement that has been made as to the operation being followed by impotence by reason of injury done to the seminal ducts. In performing the operation, it must be remembered that it is essentially a cutting one. I remind you of this, because I have known

* *British Medical Journal*, Aug. 8, 1868.

operators, before they have completed their first incision, to forget this, and begin to bore away with their index finger as if they expected to reach the staff, and afterwards the bladder, by a process of tunneling. I deprecate the reckless use of the knife just as much as I do the endeavouring to execute an operation without its sufficient employment. Such accidents as not reaching the staff at all, rupturing the urethra and pushing the bladder before you, and making an imaginary bladder between the true one and the rectum, are solely attributable to the surgeon, with the best intentions, trying to make his finger take the place of his knife. Dilatation of the prostate and neck of the bladder is only to be employed as an adjunct to a proper and sufficient incising of the parts.

In illustration of my remarks, I will briefly narrate the particulars of three cases (recorded by my dressers, Messrs. Barrow, Renner, and Porter) which, occurring almost together, afforded a good opportunity for observing many points to which I have referred.

CASE 1.—J. L., aged 15, was admitted into the Royal Infirmary with symptoms of vesical irritation. He had pain at the end of the penis, and after micturition. He passed urine frequently during the day, and had incontinence at night. A hard stone was discovered with the sound.

On Feb. 25th, 1879, I performed lateral lithotomy. The operation was unattended with any special circumstance worthy of comment. There was no hæmorrhage requiring the ligature of a vessel. The patient was removed to bed after the operation and kept within blankets until reaction set in. The operation was followed by a slight rise in temperature. On March 6th, the patient passed urine naturally. This was followed by a rigor. On April 1st the wound had healed, and the patient left the hospital.

From the history of the case it is probable that the patient had suffered from stone from childhood, and that it was the cause of the incontinence of urine.

CASE 2.—Thomas R., aged 14. This boy had suffered from symptoms of vesical irritation for over ten years. He, like the other lad, had incontinence, and pain after micturition. The pain in this case was almost entirely referred to the end of the penis, the prepuce of which was greatly elongated and hypertrophied. I never saw a case where this condition was so well marked. The patient had also occasionally passed blood, and had retention requiring catheterism. From the history which was kindly furnished me by the practitioner who sent the case to the Royal Infirmary, it appears that on one of these occasions, when either catheterism or sounding was employed, some damage had been done to the parts about the neck of the bladder, for though the patient had all the symptoms of stone, yet none could be felt.

That there was something unusual I felt sure, as it was not until the second occasion of examining him that I felt the stone and satisfied myself as to the position of the sound. On Feb. 20th, I had the patient brought into the operating theatre for the purpose of performing lithotomy.

On the staff being passed, it appeared to me to be again in the wrong position previously noted, and the calculus could only be indistinctly felt. Rather than incur the risk of doing damage by attempts to get the staff in its proper position or operate with any uncertainty, I sent the patient back to bed.

On March 5th the patient was again placed on the operating table, when, before he was secured in the lithotomy position, I passed the staff to my satisfaction. The operation of lateral lithotomy was performed without difficulty; some rather hard cicatricial tissue about the membranous portion of the urethra, probably indicating the position of an old false route, being cut through, a mulberry calculus, with a phosphatic circumference, was removed. The patient made a good recovery, and returned home on April 16th.

I have only one observation to make, and that has reference to the postponement of the operation. No one would be justified in proceeding with the operation if he had any doubt as to the position of the staff. I preferred to disappoint both myself and the large number of students assembled to witness the operation rather than incur the risk of operating with the staff in a doubtful position. Anxiety upon this point was an element of embarrassment which might have led to the abandoning of the operation after much fruitless searching, if not to the death of the patient. I preferred not to incur such a responsibility, feeling assured that an attempt on another day would most likely result in the removal of all obscurity. The course of events justified this anticipation, and we had nothing to regret.

The third case is perhaps the most interesting of the series, as it furnishes an example of the rarest form of calculus—namely, the cystin or cystic oxide, so called by Wollaston. This I believe to be the largest of the kind that has hitherto been removed by lithotomy.

CASE 3.—James D., aged 21, a dock-labourer, residing at Bootle, was admitted into the Infirmary on February 19th, 1879, suffering from symptoms of stone in the bladder. His history showed that these symptoms had extended over four years, during which period, on several occasions, he had suffered from retention requiring catheterism, and at times he had passed blood. Two years ago, a small calculus was extracted from the urethra; this was followed by some relief, but more recently he had suffered from irritability of the bladder, with pain at the end of the penis and after micturition.

Upon sounding him with a lithotrite, I found that the bladder contained a large stone measuring 2 in. by $1\frac{3}{4}$ in.; though I did not suspect that it was composed of cystin, it seemed much more flinty than any calculus I had previously felt. The size and nature of the calculus determined me to perform lithotomy. The accurate measurements I took with the lithotrite not only prepared me for meeting some difficulty in the extraction of the stone, but also indicated the appropriate means of overcoming it. There was nothing in the history or circumstances of the case which would appear to favour the formation of this variety of calculus.

On February 25th, 1879, I performed lateral lithotomy, Mr. Rushton Parker holding the staff for me. On seizing the stone, I found it, as I had anticipated, too large to extract by the ordinary prostatic incision. To avoid either tearing the prostate, which was healthy, or exercising too much dragging force on the neck of the bladder (two things which it is most important to avoid in performing lithotomy), with a straight probe-pointed knife, I divided the prostate bilaterally; this gave the required space, and the calculus was then readily extracted. After this free section of the prostate, there was very considerable hæmorrhage, not from one or two arteries which could be ligatured, but from numerous small vessels about the neck of the bladder. To meet this, I was provided with Mr. G. Buckston Browne's dilatable tampon, which answered admirably, effectually arresting all hæmorrhage. The operation was followed by considerable febrile excitement on the 26th, the temperature reaching 104°. This was the highest point attained, from which the pyrexia gradually declined, and by the seventh day the normal condition was gained. The tampon was kept in the wound for twenty-four hours. The patient made a good recovery.

The calculus weighed 1050 grains. I am indebted to Dr. Campbell Brown for the following analysis. "I find that it consists of cystin or cystic oxide, with a small quantity of ammonia, magnesium-phosphate, and only a trace of calcium-phosphate. The hard, waxy, crystalline granules are well marked, and belong to the hexagonal system of crystallisation: by solution and recrystallisation the hexagonal plates are obtained perfect." The urine has been examined several times since the operation, but no cystin has been detected.

On reference, I find that hitherto the largest cystin calculus extant is the one in University College Hospital Museum. It was removed by Liston, and described by Dr. Bence Jones in the *Medico-Chirurgical Transactions* for 1840. It weighed over 850 grains. The Museum of the Royal College of Surgeons, out of a total of about one thousand calculi, I believe, contains only five specimens of the cystin variety.

Before concluding my remarks on lithotomy, I would refer to a recent paper on this subject by Mr. Furneaux Jordan,[*] in which he briefly epitomizes a mode of operating he advocates in the following words. "Open the membranous urethra on a grooved staff; put a strong simple concealed bistoury into the bladder; project the blade *upwards*, and draw out the bistoury; put the finger into the bladder, feel the stone, and extract as usual." I have not had any

---

[*] "On a Method of Performing Lithotomy."—*British Medical Journal*, Jan. 24, 1880.

experience of this mode of operating; I cannot, however, see that it possesses any advantage over what appears to me to be the simplest and most generally applicable proceeding—namely, the lateral operation.

TWENTY-THIRD LECTURE.

LITHOTRITY—LITHOLAPAXY—STONE IN FEMALES—TREATMENT.

The other plan of removing a stone is by lithotrity, by which it is broken up into fragments capable of being expelled during the act of micturition, or withdrawn by some suitable contrivance.

The conditions favourable to this operation may be generally summed up as follows: an adult and otherwise healthy person, a stone of moderate size and hardness, a normally proportioned urethra, and a fairly tolerant bladder. These are the conditions we most desire to secure, and the nearer they are approached the greater is the likelihood of a favourable result.

All cases do not at first come up to this standard, but, by preparatory treatment, complications may be removed or so mitigated as to bring many of them within the limits of lithotrity.

I do not attempt further to define the boundaries of this operation, as the experience of the present day is, I think, likely to extend rather than to curtail them. I shall therefore proceed to notice a few points in connection with the performance of lithotrity.

First, as to anæsthetics. A more extended experience has led to my universally adopting them, not only in lithotrity, but in all operations on the urinary organs where pain is likely to be occasioned. In my earlier cases my hesitation to employ anæsthetics was not based on the ground of the risk of damage being done to the bladder which a conscious patient might prevent. Such an objection would apply with about equal force to almost every operation in surgery. It was rather because I did not find the operation to be one that taxed much the endurance or fortitude of the patient, as the time of a sitting then never exceeded two or three minutes. A great change has come about in this latter respect, and I now employ an anæsthetic in lithotrity as invariably as I do in all other operations in surgery where it is admissible.

Next, as to the selection of a lithotrite. None but those who, like myself, have had to operate with the old rack-and-pinion instrument can fully appreciate—I was going to say—the luxury of using the modern instrument, the joint production of Sir Henry Thompson and Messrs. Weiss.* It combines every element necessary for the most delicate manipulation and the application of the requisite force.

Further, I would lay stress on carefully studying the proper position for each patient during the performance of the operation. As Sir Henry Thompson

* Whether, in the face of the very perfect instrument we at present possess, the lithotrite is capable of being further improved is doubtful. I cannot, however, in justice to Dr. G. C. Duncan, of Montreal, refrain from noticing a most ingenious application of the drill which he has adapted for this purpose. He has kindly shown his instrument to me and my class; but, inasmuch as it has not yet been submitted to the actual test of experience, I cannot say more. Dr. Duncan is the author of a suggestive article on Litholysis, which appeared in the *Edinburgh Medical Journal* of May, 1877.

points out, there is an "area" in every bladder in which to operate. You want to make that area correspond with the position taken up by the lithotrite as it lies in the bladder. I have often noticed how one pillow more or less, placed under the pelvis, makes all the difference in the facility with which the fragments are caught at successive sittings.

I prefer operating with some urine in the bladder, and I therefore instruct patients to retain their urine if possible, or, if this is not sufficient, I inject the necessary quantity of tepid water.

Up to within quite recently the rule in lithotrity has been for the operator to limit the use of the lithotrite to something like a couple of minutes, repeating these operations, or "sittings," as they have been called, until the stone has been completely broken up and discharged from the bladder, the intervals between each sitting varying according to circumstances. In the case of small stones, of course, a single sitting has sufficed without exceeding the limit of time I have mentioned. Though three minutes seem a short period—especially when not measured by the watch—yet you will see a surgeon who is accustomed to use the lithotrite pick up and crush several fragments at a sitting in considerably less time than this. Much of this adroitness is due to the operator having previously carefully considered his bearings, and the best position in which to place the patient, with a view of limiting as far as possible his area of action.

After each sitting the practice of surgeons differed; some endeavoured, by successive introductions of the lithotrite, to bring away as much of the *débris* as possible, or to withdraw the fragments by a suction apparatus fitted to a large-eyed catheter, such as Clover's; whilst others preferred keeping their patient in bed for a short time after each sitting, and not in any way hastening the discharge of the pieces until they had been sufficiently triturated to escape with the urine during micturition. I adopted the latter practice, for the reason I expressed in a discussion which followed my exhibiting, for the first time in England, Dr. Bigelow's apparatus [*]—namely, that though the evacuators then employed for this purpose, including those of Sir Philip Crampton and Mr. Clover, were excellent in principle, yet they were deficient in design, inasmuch as they removed but little else of the detritus than that which would readily escape during micturition, the larger fragments, which were those most likely to excite irritation, being left behind.

That the retention of these fragments in the bladder, with the distress they occasioned in their passage along the urethra, was a frequent cause of a fatal result in cases otherwise favourable, there cannot be the least doubt, and such being the case, we can see the force of the recommendation that the best mode of treating the acute form of cystitis, following a "sitting," was the further and speedy breaking up of the fragments left behind. In view of this position, Dr. Bigelow, of Boston, Professor of Surgery at the Harvard University, has recently submitted to the profession an operation

---

[*] "Report of the Proceedings in the Surgical Section of the British Medical Association," Annual Meeting at Bath.—*British Medical Journal*, Aug. 24, 1878.

which he has designated "litholapaxy," planned with the object of doing away with that serious cause of mortality to which I have referred.

During my stay in Boston, in 1878, I enjoyed the privilege of seeing Dr. Bigelow operate, and of receiving from him a full explanation of the various details employed. Dr. Bigelow advocates the complete breaking up of the stone at one sitting, and the removal of the fragments by an efficient aspirator, contending that the intolerance of the bladder to the presence of the lithotrite has been very unjustly exaggerated, and that practically there is no time-limit to the skilful use of the instrument. In the case I saw, Dr. Bigelow occupied one hour and nineteen minutes in the breaking up and removal of a uric acid stone. The patient was under the influence of ether during the whole time, and made an excellent recovery. I saw him up to the fourth day, when convalescence was pronounced. A careful consideration of Dr. Bigelow's work will undoubtedly lead to the more general adoption of the views he has advocated as to the treatment of stone by crushing; at present litholapaxy must be considered as being submitted to the most crucial of all tests—namely, that of practical experience. As illustrating this mode of operating, I will quote a recently published case of my own.*

J. A., a ship-keeper, aged 68, was admitted into the Liverpool Royal Infirmary on April 22nd, 1879. Eight years before he suffered from symptoms of stone, for which he was cut, and a calculus was removed. Within a year he was again cut at

* *The Lancet*, Oct. 25, 1879.

another hospital, and two calculi were successfully removed. During the year previous to admission a stone was removed by lithotrity, and at the commencement of the present year the old symptoms again returned, when for the first time he came under my notice.

On admission he was suffering from frequent micturition, pain at the end of the penis, and occasional hæmaturia. The urine was purulent and deposited phosphates in abundance. On examining him under ether with a lithotrite, a round stone of about an inch in diameter was felt.

On April 25th, the patient being placed under ether, I broke up the stone, which was phosphatic, and evacuated the *débris* after the manner practised by Professor Bigelow. The symptoms were at once relieved, and the cystitis from which the patient had so long suffered gradually abated under treatment.

Though the calculus was most completely broken up, there were no signs of hæmorrhage, either during the time of operation or afterwards. He left the Infirmary on May 15th, 1879, passing urine normally and free from all signs of calculus. Whether, after such a history as that mentioned, he will continue free from further calculus formations remains to be seen.

After so extended an experience of stone and its treatment as this patient had undergone, it is not only interesting but proper to record his observation, "that the new plan was the easiest way he had had of getting rid of a stone."

An increasing experience in this mode of operating leads me to believe that Dr. Bigelow's practice will materially add to our resources for the treatment of stone.

The chronic form of cystitis which sometimes supervenes after lithotrity, indicated by an irritable state of the bladder and unhealthy muco-purulent urine, is generally due to some of the latter being constantly retained; watchful care should always be taken to guard against this, and, when discovered, to prevent its ill consequences by catheterism and irrigation, as described in a previous lecture.

Though, for obvious reasons, lithotrity is best suited to adults, it is not necessarily restricted to them. At one crushing I removed from a boy, eight years of age, a small uric acid calculus, and recently, in two sittings, I completely broke up a similar, though larger, calculus, in a young gentleman aged sixteen years.

It is sometimes remarkable what large fragments may be voided through the urethra after lithotrity. I have some fragments weighing as much as twenty grains which were passed, with others only a little less, in a case I attended with Mr. Gaskell, of St. Helens, where, at seven sittings, I crushed a uric acid calculus two inches in diameter. The patient was forty-two years of age, and made a good recovery. He was possessed of a urethra admirably adapted for such an operation. I need hardly add, as a comment upon this case, that in performing lithotrity our object should be to pulverize the fragments as completely as possible.

Litholysis, or the dissolving of calculi, may sometimes be advantageously combined with lithotrity.

The solvent action of nitric acid upon the *débris* of a phosphatic calculus was well illustrated in the following case, which in many respects is similar to one recorded by the late Mr. Southam, and alluded to by Dr. W. Roberts.[*]

[*] *On Urinary and Renal Diseases*, by Dr. W. Roberts, 2nd Edition.

William C., a police officer, aged 45, was admitted into the Liverpool Royal Infirmary, under my care, on March 9th, 1875. The patient had recently been under the observation of my friend Dr. J. S. Clarke, who, having diagnosed the existence of a stone in the bladder, advised his admission.

Upon examination, I found a single stone, rounded and having a diameter of two inches and a quarter. The urine was alkaline and contained pus. Though the stone was large, I deemed the case not unfavourable for lithotrity. The patient required some preparatory treatment, and I was not able to commence lithotrity till April 5th. I repeated the crushings during April and May twelve times.

The patient never experienced any unfavourable symptom, and a very considerable quantity of broken-up stone was passed. As in Mr. Southam's case, to which I have alluded, it appeared to me that fresh phosphatic depositions were taking place almost as fast as the others were removed. I resolved, therefore, to use nitric acid injections.

On May 27th (two days after a crushing), I had the urine of twenty-four hours collected. This was analysed by Dr. J. Campbell Brown, who gave me the following report. The quantity of urine submitted to analysis was sixty ounces; this was found to contain 74·029 grains of phosphoric acid as alkaline earthy phosphates; 250·516 grains of phosphoric acid as alkaline phosphates; in all, 324·545 grains of phosphoric acid passed in twenty-four hours. At the termination of this period, I injected into the bladder half a pint of tepid water, with two drachms of diluted nitric acid.

All the urine passed during the subsequent twenty-four hours was kept and analysed by Dr. Brown, with the following result. The quantity of urine submitted to analysis was seventy-eight ounces; this was found to contain 96·237 grains of phosphoric acid as alkaline earthy phosphates; 461·94 grains of phosphoric acid as alkaline phosphates; total, 558·177 grains of phosphoric acid passed in twenty-four hours.

Thus it appears that not only was a larger quantity of phosphate of lime and magnesia dissolved in the urine after treatment with nitric acid, but there was a still greater increase in the quantity of phosphoric acid passed in the form of alkaline phosphates. It is to be noticed that the quantity of urine on the day after treatment was greater than on the day before; and if, instead of estimating the total phosphates, we estimate the percentage of phosphates in the urine, we find that the percentage of alkaline earthy phosphates was very nearly the same after as before treatment—namely, increased from 0·272 to 0·282, and that the percentage of alkaline phosphates was increased from 0·95 to 1·35. But the absolute increase of phosphates is much more marked.

On nine subsequent occasions the lithotrite was employed, and, on every second or third day afterwards, the bladder was injected as before with nitric acid in tepid water. Further observations, though made in a rougher manner, showed the increase of the alkaline phosphates after each injection; it was also noticed that the fragments passed were much more finely triturated than previously.

Under this treatment the patient made a good recovery, and left the Infirmary quite well on June 21st. The total quantity of broken-up stone collected weighed four drachms. The use of the acid appeared to me to stop any further deposition of phosphates, and to facilitate the removal of the pieces as they were broken up by the lithotrite. I saw this patient two years after he left the Infirmary. He is strong and well and has had no recurrence of stone.

Dr. Roberts concludes his observations on Mr. Southam's case with the remark: "This method is evidently capable of wider application than is now made of it by surgeons."*

* *Op. cit.*, p. 309.

Stone in females is of less frequent occurrence as compared with the opposite sex; a circumstance which is probably due to the shortness of the urethra favouring the escape of a calculus at its earliest formation.

For the treatment of stone in females, rapid dilatation of the urethra and extraction are usually employed where the calculus is small. When large, it should first be broken up by the lithotrite, and then removed piecemeal.

I was much disappointed in the case of a child six years of age, upon whom I operated in 1876, to find that after she left the Infirmary, apparently well, incontinence returned. The patient being placed under ether, dilatation was readily accomplished by means of Otis's dilator for stricture of the male urethra, which answered this purpose exceedingly well. The stone was almost round, and hardly an inch in its broadest diameter. On a future occasion, in stones of this size in female children, though in the case referred to I do not think I exceeded the limits which some authors give, I shall first use the lithotrite, rather than run the risk of incontinence occurring.

As indicating the extent to which the female urethra may with safety be dilated for the purpose either of exploration or of extracting a stone, the following passage from a recent author may be quoted.

" Simon, of Heidelberg, has made dilatation of the female urethra a proceeding applicable with scientific accuracy. The urethra can be dilated to a diameter of 1·9 to 2 centimetres, or ¾ inch, in women over twenty years of age ; to 1·8 centimetres, or rather more than

1¼ inch, in those between fifteen to twenty; and to 1·5 centimetres, or ⅝ inch, in those between five and eleven. Under twenty years of age these measurements may, in case of need, be exceeded by 2 or 3 millimetres. In no case does incontinence of urine result. Simon's statements have now been verified by general experience. Hence, since the average diameter of a man's right index finger at its thickest part is about ¾ inch (1·8 cm.), and of his little finger ⅝ inch (1·5 cm), it may be stated that we can safely dilate the adult urethra so as to admit the index finger, and the child's so as to admit the little finger."[*]

In the case of very large stones, which cannot be included within the grip of the lithotrite, removal has been effected by supra-pubic or vaginal lithotomy; the latter plan is the one which on all grounds is to be preferred. This operation consists in opening the vaginal wall of the bladder by a median incision, and, after extracting the stone by forceps, reuniting the edges of the wound by sutures, as is done for vesico-vaginal fistula. In a very interesting paper by Dr. Warren,[†] this operation is discussed and illustrated. The worst danger likely to follow it is incontinence from failure of the wound to unite, a remediable condition, and therefore very different from the incontinence that follows over-distension of, or incision into, the female urethra. Dr. Galabin has recently recorded a case where by this operation he removed

---

[*] " On the Operation for Stone in the Female Bladder," by A. Ogston, M.D.—*Edin. Med. Journal*, July, 1879.

[†] " On Vaginal Lithotomy," by J. C. Warren, M.D.—*Boston Medical and Surgical Journal*, July 20, 1876.

twelve large calculi and about fifty small ones from the bladder of a woman aged 61. The wound was closed by silkworm gut sutures, and at the end of ten days union was complete.*

The female urethra is occasionally made the resting-place for a stone. Some time ago I assisted Dr. Lyster to remove a uric acid calculus lodged in a sulcus in the floor of the canal. The stone was of an oval shape, and measured one inch and a half in length and three-fourths in breadth. Removal was effected by rapidly dilating the urethra with Weiss' instrument, and extracting with forceps. The patient recovered without a bad symptom.

Phosphatic concretion on the walls of the bladder is met with both in males and females. A well-marked instance of this I recently saw in consultation with Dr. Matthew Hill, of Bootle, where we found the mucous membrane of the bladder in a female deeply encrusted with this deposit. It was necessary to dilate the urethra rapidly, and then with the finger to peel off, as far as possible, the concretion. This was followed up by injections of a weak solution of nitric acid, under which treatment the patient progressed most satisfactorily.

* *Obstet. Society's Trans.*, April 7, 1880.

TWENTY-FOURTH LECTURE.

Injuries to the Bladder.

Rupture of the bladder is an accident which until recently has been passed over with comparatively little notice, from which we may infer, not that those who have preceded us were less capable of determining the nature of the lesion, but that their means of dealing with it were very limited. It is not surprising, therefore, to find, with certain improved appliances in surgery which will for ever render the present age memorable, that further attention has been devoted to this subject. And these improved means of applying surgery consist in the introduction of anæsthetics, and the exploration with a far greater degree of safety of the various cavities of the body by the employment of antiseptics.

It will hardly be necessary for me to refer at length to the circumstances which may render the bladder structurally incapable of holding its contents, so as to allow the urine to escape either into the cavity of the peritoneum or amongst the adjacent tissues. The causes of this condition include violence from without, such as crushes, or from within, as penetration by fractured bones; or, more rarely, excessive muscular action;[*] we have, likewise, similar effects produced by bullets and other projectiles. In the examination of these cases we shall have no difficulty in recognising two distinct varieties : (1) Where the cavity of the peritoneum is opened into; and (2) extra-peritoneal, where the rupture is in a part of the bladder which is beyond its peritoneal investment. The difference is necessarily this, that in the former variety the urine passes directly into the peritoneal cavity, whilst in the latter it infiltrates the tissues around the bladder, and produces consequences in every respect similar to those observed in parts which are more superficial, where the effects caused by urinous infiltration can be seen.

It is important to analyse these two conditions as closely as possible, with the view of narrowing the principles of treatment applicable to each, and with this object I shall refer to the records of clinical experience we possess, such as, amongst others, the paper by Mr. Heath, and the not less interesting and important discussion which followed it.[†]

Taking those cases where the rupture has extended into the peritoneal cavity, I cannot find any evidence to warrant the belief that life has ever been saved without the assistance of surgery. Though the examination of the laceration, in cases where life has been prolonged for some days, affords evidence that nature, ever mindful of her powers of conservation, has been occupied in attempts to repair the damage, I cannot

---

[*] "Case of Rupture of the Bladder, apparently without External Violence or Stricture," by Dr. Macewen.—*The Lancet*, Sept. 27, 1873. Also see page 39.

[†] *Royal Medical and Chirurgical Society*, Feb. 25, 1879.

conclude that, unaided, these attempts have ever been successful.* Whereas, on the other hand, the few instances that have recovered from this most serious injury can be directly traced to the aid that nature has received from art. It will be proper now to see under what circumstances surgery has contributed to the preservation of life in this class of cases.

In the first place, there are grounds for concluding that catheterism alone has been effective in bringing about a favourable result; the cases of Mr. Chaldicote and of Dr. Thorp (referred to in Mr. Heath's paper) are instances of this. In these there can be no doubt that large quantities of urine were drawn off from the peritoneal cavity, which, if allowed to remain, it is reasonable to suppose would have induced fatal consequences. In a recent communication of great value,† Dr. Macdougall brings forward two cases of recovery after rupture of the bladder where catheterism afforded important aid in the successful treatment. In one, aspiration of the abdominal cavity was also employed. The view taken that the use of catheterism in these and other instances determined the satisfactory result

---

* "Nearly all the cases of this injury (rupture of the urinary bladder) hitherto reported have been fatal; yet in a few of these, where the patients lived several days after its occurrence, on examining the injured parts, processes were observed apparently tending to their recovery; these processes consisted in the confinement of the effused urine within a sac bounded by false membrane, the product of inflammation in the surrounding tissues, and they have been observed, not only in cases where the urine has been effused into the cellular tissue on the outside of the peritoneum, but also in the instances of its effusion into the peritoneal cavity."—On Rupture of the Ureter, by E. Stanley, F.R.S., *Medico-Chirurgical Transactions*, vol. xxvii.

† "On Rupture of the Urinary Bladder, and its Surgical Treatment."— *Edinburgh Medical Journal*, Jan., 1877.

is strengthened by a reference to some recorded cases which terminated fatally. There is good reason for believing that the peritoneal cavity is more tolerant of the presence of healthy urine than at first sight we might be inclined to suppose. In fact, I would go further and say, that in these cases the fatal peritonitis set up, is not due so much to the entrance of healthy urine within the limits of an uninjured peritoneum, as to the decomposition of urine which follows its confinement by, or even contact with, tissues more or less disintegrated by violence. Menzel's experiments * are confirmatory of this view, as they demonstrate that healthy urine does not in itself cause destruction, and that its effects on the tissues are innocuous so long as an escape is provided for it. Clinically, we have illustrations of Menzel's important experiments in cases such as the following, where rupture of the bladder of a man was caused by a blow in a pugilistic encounter.†

"At the time of the accident a sensation of something having given way was experienced by the patient. He walked home, a distance of two miles, when he was seen by Mr. Robinson, his medical attendant, who drew off twenty ounces of bloody urine, and continued to do so twice a day for three days until he was admitted into the Leicester Infirmary. On his admission, Nov. 20, 1871, he presented the following signs:—He was able to walk without assistance; countenance rather anxious; pulse 80, full; skin cold; complained of a sensation of weight in the hypogastric region, but no tenderness on pressure. On percussion over the abdomen an increased area of dulness was

---

* *Wien. Med. Wochenschrift*, Nos. 81–85, 1869.
† *Medical Times and Gazette*, Sept. 28, 1872.

detected, and on palpation a distinct sensation of fluctuation. A silver catheter was introduced, and thirty ounces of clear urine drawn off, the abdominal dulness and fluctuation entirely disappearing. A gum-elastic catheter was subsequently passed night and morning. The patient continued much in the same state up to the evening of December 1st, when, without any premonitory symptoms, he was seized with a severe attack of convulsions, rapidly followed by coma. Every effort was made to rouse him; a catheter was immediately introduced, and about twenty ounces of urine drawn off, consciousness returning in about an hour. On the following day he was again attacked in a similar way, and on December 3rd he was seized with a more violent form of convulsions, and sank in three hours in a comatose state.

"*Post-mortem examination.*—Slight bruise over the scrotum. On opening the peritoneal cavity, about four pints of clear fluid welled up through the incision. The bladder was found contracted, and a laceration of its posterior surface to the length of two inches was detected extending in an oblique direction. In other respects the organs presented no morbid changes.

"*Remarks.*—This case presents the following points of interest:—First, the power of locomotion after so serious an accident, the patient having walked to his home, a distance of two miles, immediately after the occurrence. Secondly, the length of time he survived—sixteen clear days,—the average duration of life in these cases being from three to seven days. Thirdly, the absence of all signs of peritonitis. Fourthly, could the death of the patient be attributed to peritoneal absorption?"

In this very interesting case there can be no doubt that the patient was almost within the limits of recovery. He was poisoned, I believe, as is suggested, by his own urea, before nature had time to provide some compensating action by which it could have been got rid of.

Further, such instances seem to show that, if no extensive damage is done to the structures of the abdomen, if the urine that finds its way into the cavity of the peritoneum is not largely contaminated with blood or other source of putrefaction, and if escape, as by catheterism, is provided for urine so effused, serious peritonitis need not necessarily be provoked.

In the majority of cases of intra-peritoneal rupture of the bladder, one or other, or all, of these conditions are usually absent. The urine, under such circumstances largely mixed with blood, decomposes and gives rise to that rapid and destructive inflammation which, in spite of all treatment, brings these cases to a fatal termination in the course of a few days. To meet cases such as these, obviously hopeless if left to themselves, it has been proposed to perform abdominal section antiseptically, and to close the opening in the bladder with a continuous suture. Illustrations of this practice have been recorded both by Mr. Heath and Mr. Willett, but I cannot say that the results so far obtained are encouraging, though, as Mr. Holmes observed in the discussion to which reference has been made, in both these instances the fatal issue was due to the giving way of sutures, so that neither could be quoted against the probable efficiency of the operation.* Still there is much to be said in favour of the principle upon which such a proceeding is based, and it is possible that by improved modes of opening

* *Royal Medical and Chirurgical Society*, Feb. 25, 1879.

the abdomen, in conjunction with antiseptics, this means may add other recoveries to those which have been obtained by catheterism of the abdominal cavity.

Where the bladder is ruptured external to the peritoneal cavity, and a vent is made for the escape of extravasated urine, recovery may occur in spite of the severe nature of the lesion. As bearing upon this, I will cite a case which happened in my own practice at the Northern Hospital.

A middle-aged man was admitted under my care, in 1866, for injuries received by the fall of some earth whilst excavating for some dock extensions. His pelvis and abdomen were much crushed. There was a fracture of the right ilium, with considerable bruising of the adjacent soft parts. On passing a catheter the bladder appeared to be contracted on the end of the instrument, and nothing but a few drachms of blood-stained fluid escaped. There was a suspicion then that the bladder had been ruptured. The catheter was retained. In the evening I saw him with my colleague, Dr. Lowndes. He was much collapsed; no urine had escaped or could be withdrawn. The condition of the bladder, as felt by the catheter, was too suspicious to permit of a conclusion being drawn that the shock of the injury had occasioned a suppression on the part of the kidneys. The perinæum externally was somewhat tumid, but not discoloured; on introducing the finger into the rectum, an unnatural distension in front of the rectum could be felt which quite concealed the boundaries of the prostate. Under these circumstances, we thought it better to make an incision in the median line in the direction of the prostate. This I did, and gave vent to a large mass of clots and fluid, which I believe was urine. On introducing the finger into the wound, the prostate appeared to be dissected off the parts beneath, and a depression could be felt on its under surface which proved to be the termination of a laceration. Through this deep and extensive wound blood-stained fluid continued to flow during the six days he lived. The view taken at the time was that it was a laceration at the neck of the bladder. So far as the injury was concerned the patient might have pulled through, but the other injuries added to this one were too much for him, and he succumbed of exhaustion on the seventh day. At a post-mortem examination the prostate and neck of the bladder were completely separated from the parts beneath, and there was an extensive rent commencing an inch behind the prostate and reaching forwards through it. The wound did not communicate with the peritoneal cavity, and there were no signs of peritonitis. There was a comminuted fracture of the right ilium, the fracture passing down within the brim of the pelvis.

The practice adopted in this case would have probably saved the patient's life had there been no other complication, as the incision provided a channel of escape for the urine as direct and free as I could have wished. Here the only reason we had for supposing that the rupture was within reach, in addition to the evidence that the catheter and the nature of the injury afforded, was the tumefaction of the perinæum and the fulness felt in front of the bowel on introducing the finger into the rectum. This, coupled with the fact that no urine could be obtained by the catheter, justified the conclusion we arrived at and the practice we adopted. It will be observed that in my case a median perinæal incision to the neck of the bladder was made. A lateral lithotomy has, under somewhat similar circumstances, been substituted for this procedure.* The median incision, I think, is to be preferred, as it is less likely to be attended with such hæmorrhage as would require the use of some controlling apparatus in the wound, similar to the "umbrella tent" sometimes employed after lithotomy, the use of which would be inconvenient in parts swollen and contused by the violence producing the rupture.

It will be seen, from a consideration of the cases of ruptured bladder, that I have laid stress on certain principles of treatment as being applicable to the two varieties of this lesion. I admit that, owing to the damage done to the adjacent tissue and the large amount of ecchymosis, it is in practice no easy task to draw such a distinction; it is, however, only by a careful scrutiny of such cases, unencouraging as they are, that we can hope, by the employment of the means I have indicated, to secure less fatal results than have hitherto attended rupture of the bladder.†

Penetration of the bladder by sharp instruments, or by bullets and other projectiles, is occasionally met with in civil as well as in military practice. An unusual case of this kind is recorded by Mr. Treves as occurring in the practice of Mr. Couper at the London Hospital.‡

A seaman, aged 23, was admitted under the care of Mr. Couper on April 26, with a small incised wound of the left buttock. He had been stabbed, in a quarrel, with a long sailor's

---

* Dr. Mason's Case. (Mr. Heath's paper.)

† A careful criticism of some of the views I have advanced will be found in a Clinical Lecture by Mr. Henry Morris, *Medical Times and Gazette*, Nov. 29, 1879.

‡ *Medical Times and Gazette*, June 14, 1879.

knife. He walked into the receiving room, but with difficulty; was blanched, and in a condition approaching collapse. From the condition of his trousers it was evident that he had lost a great deal of blood. The wound was about one inch long, was clean cut, and situated exactly in the middle of the left buttock. The finger introduced into the wound passed for some depth in the direction of the great sciatic notch, and that notch, indeed, could be felt. There was a little venous oozing from the wound, which was immediately checked by pressure. The wound was treated antiseptically. The patient complained of no pain other than that immediately about the wound. Shortly after admission he passed his water; it was clear, and contained no blood.

Next day patient appeared quite comfortable, made no complaint, had no difficulty with his water, which was passed at usual intervals and was always normal.

April 28.—Patient vomited several times; complained of pain over abdomen; became restless. Temperature at night 104°.

April 29.—Vomiting continued. Patient very restless and feverish; much pain complained of over hypogastric region. In the afternoon the patient tore off all the antiseptic dressings (they had been renewed daily). The wound looked well and was free from discharge.

Symptoms of acute peritonitis now became more apparent, and death took place on May 1.

Throughout the whole case the patient never had any trouble with his bladder.

At the post-mortem examination the knife was found to have taken the following course:—It had penetrated the gluteal muscles, divided a part of the great sacro-sciatic ligament, and passed through the small sacro-sciatic notch, completely dividing the pudic artery and nerve and one vein. Each end of the pudic artery was perfectly closed by a little clot. The knife had then entered the bladder at its lower part and close to the trigone, making a wound large enough to admit the tip of the forefinger. There was diffuse suppuration of the cellular tissue of the pelvis, and general acute peritonitis.

The case is interesting as illustrating a complication which, from the symptoms, there were no grounds for suspecting, post-mortem examination alone revealing it. In cases of incised wounds, where there is a suspicion or evidence that the bladder is wounded, it is probably the best practice to allow the urine to drain off by a catheter, passed either through the wound, should it be in a dependent part of the bladder, or along the urethra only, whichever way it flowed the more freely, rather than to permit of it collecting and being discharged by the expulsive power of the bladder in the ordinary act of micturition. In wounds of the anterior wall of the bladder, I feel sure, from the experience of supra-pubic incisions made for lithotomy, for stricture, and suicidally, that the best plan is to accurately close the opening in the bladder by deep continuous sutures, as we should do for a wounded intestine, and subsequently, by superficial sutures, to bring together the opening in the skin, allowing the urine to drain off by a retained catheter.

On gunshot wounds involving the bladder I do not intend to dwell at any length, as their discussion belongs to a branch of surgery of which, as civilians, we see but little. A series of most interesting cases of this kind will be found in a work which illustrates every variety of wound involving the viscus inflicted in warfare.*

* *Medical and Surgical History of the War of the Rebellion, U.S.A.*

In the management of these cases, in addition to the employment of the general principles applicable to the treatment of wounds, there are one or two points which must not be lost sight of. By incision or by catheterism, the most direct escape should be provided for the urine, as its collection in the locality of extravasated blood or damaged tissue is a fruitful cause of destructive inflammation. Further, a foreign body which has found its way directly into the bladder, or indirectly by ulceration or sloughing, may serve as a nucleus for the formation of stone; of this we have instances, in the work just referred to, of concretions forming on arrow-heads, such as are used by the Indians, or bullets and other projectiles, on pieces of bone, or on the *débris* of a fractured pelvis which have made their way into the bladder. Portions of clothing, buttons, and other articles of dress, in like manner, have become covered with phosphatic concretions.

It is asserted that lacerations of the bladder may occur without extravasation of urine ensuing as a consequence. I do not see what positive evidence of this can be afforded; the nature of the injury and hæmaturia may possibly suggest it, but this is about all that can be said. If I were to suspect that a slight rupture had taken place, which by some means or other—such, for instance, as the presence of a clot in the wound, or the exudation of inflammatory material—had become occluded, I should not feel disposed either to pass a catheter, provided there was no retention, or to retain one. I would sooner trust to nature completing a task she had commenced so well, aiding her, perhaps, in keeping the parts quiet and the skin active, as we do with a ruptured intestine, by the internal administration of opium. The action of the bladder may be said to be permanently arrested in the majority of cases of fracture of the spine. I have referred to this elsewhere, in my remarks on the importance of carefully attending to the removal of the urine and the general management of the bladder in these injuries.

The power of the patient to voluntarily expel his urine is occasionally temporarily suspended in connection with other injuries to the trunk. Retention of urine not unfrequently follows concussion of the spine. Power usually returns in the course of forty-eight hours, and, so far as this symptom is concerned, nothing further than regular catheterism is required. In courts of law imperfect power in micturition is sometimes referred to as a symptom of spinal concussion. When it immediately follows an injury and disappears as I have already mentioned, it need not create apprehension, as it is only in keeping with the other signs of nerve shock. When it occurs after a lapse of time, and consequent upon an injury, its import is unquestionably grave, as it is probably connected with destructive structural changes in the nerve centre which controls the action of the bladder.

I have seen retention of urine occur in connection with violent contusions, such as crushes involving the abdominal muscles. Their inability to co-operate with the bladder in the expulsion of urine, without causing severe pain, is sufficient to explain this.

TWENTY-FIFTH LECTURE.

Injuries to the Ureter and Kidney—Hernia of the Kidney.

Rupture of the ureter, or its division by an incised wound, is an accident of comparatively rare occurrence. I have once seen it, in connection with a severe contusion of the loins, in a patient who, whilst intoxicated, was badly crushed by a heavy waggon which was supposed to have passed over him in the dark. I think it more probable, from the nature of the injuries, which included a fracture of the last two ribs on the left side, a contusion of the adjacent parts, and a rupture of the ureter close to the left kidney, as well as from a description of the circumstances under which the man was found, that he had been squeezed between the waggon and a large stone. The patient died on the day following his admission to the Northern Hospital without perfectly recovering consciousness. Some blood-stained urine was removed by the catheter, but beyond this there was nothing requiring any special comment as indicating the full extent of the damage.

Mr. Stanley[*] has recorded two instances of rupture of the ureter. In the first the patient recovered, and there was therefore no opportunity of verifying the

---

[*] *Medico-Chirurgical Transactions*, vol. xxvii.

diagnosis which had been made. In both of these cases the most prominent feature was the collection of fluid in the cellular tissue behind the peritoneum, which had to be removed by tapping. In the second case the patient died in the tenth week after the injury, and post-mortem examination revealed the correctness of the opinion formed. It is curious to observe that in each instance, though the fluid removed had a resemblance to urine, yet it was deficient in some of its characteristics, as shown by chemical examination. This is a point to which I shall again refer. Mr. Poland* has also placed on record a similar case where, consequent on a violent crush between the platform of a railway station and a moving train, in addition to other injuries, the ureter was ruptured, the patient surviving 135 hours. As in my own case, the damage to the ureter was obscured by the extent of the other abdominal injuries. At the examination after death it was found that the right ureter was torn across just below the pelvis of the kidney. Following immediately upon Mr. Poland's case, in the same volume of reports, is an article by Dr. Moxon, "On two Cases of Thrombosis of the Renal Vessels, through injury to the Lumbar Spine; with general Remarks on Thrombosis." This paper is partly based on the condition of the kidneys as observed in the case of Mr. Poland, to which reference has been made, where these organs were found with the vessels blocked by ante-mortem clots, and showing other indications of the violence to which they had been exposed. It is almost impossible to imagine that a

* *Guy's Hospital Reports*, 1869.

rupture of the ureter could be effected without the greatest amount of violence short only of immediate destruction of life. Even supposing that the requisite force could be limited to the ureter, and continued as by a pull or pressure until it snapped across, one cannot see how such a tube could be so broken without the kidney to which it is attached being, to some extent, structurally damaged. Is it not possible that the damaged condition of the kidney which must be associated with a tear of the ureter, is sufficient to account for the supposed alteration of the excretion which has been observed in cases already referred to (Mr. Stanley's), and in one presently to be mentioned (Mr. Holmes'), where the only doubt as to the nature of large quantities of fluid that escaped from the neighbourhood of a ureter which was believed to have been incised, arose from the fact that examination failed to detect the presence of those salts which healthy urine usually contains? In looking over the analyses of these fluids, they appear to me to be very similar to urine found in another condition of damaged kidneys —namely, in certain forms of Bright's disease, where these organs are scarcely anything but aqueous percolators. I think it exceedingly probable that thrombosis of the renal vessels commonly occurs in connection with rupture of the ureter, and that as a consequence of this the excretion of the damaged kidney is little else than water. If this were not so, how happened it in Mr. Stanley's cases that the escape of urine into the cellular tissue from the ruptured ureters was not followed by those signs which usually accompany extravasation? Is it not reasonable to believe that in these instances the escaped urine failed to arouse active inflammatory changes, because of the withdrawal from it of those salts, as urea, which produce ammoniacal decomposition?

Mr. Timothy Holmes, in a communication "On direct Wounds of the Ureter,"[*] narrates a case where the ureter had been opened by a stab-wound in the back. The injury was followed by the discharge from the wound of immense quantities of clear fluid, which, though differing from urine in composition, was most probably the excretion of the kidney in some altered form. The case is one of much interest, and though not free from ambiguity, as, by reason of the recovery of the patient, the diagnosis was not confirmed by anatomical examination, yet the conclusion arrived at by a critical analysis of other explanations which were suggested is almost irresistible. Might not, I repeat, the altered condition of the urine which was discharged so abundantly through the loins be due to a similarly infarcted state of the bloodvessels of the kidney—as was observed in Mr. Poland's case—consequent on the proximity of the wound to this organ? Is it too Utopian to suggest that nature provides for a lacerated ureter by the induction of such changes in the corresponding kidney as will render its excretion the least hurtful to the tissues with which it may come in contact? I cannot help thinking, from the cases I have just referred to, in addition to another recorded in my lecture on extra-

[*] *Medico-Chirurgical Transactions*, vol. lx.

vasation, where a strong probability is furnished that urine which does not contain urea is incapable of producing distinctive changes in the tissues surrounding it, that we have in the view I have advocated the explanation that Mr. Holmes considers wanting in the following remark on his own case: "If it could be shown that a wound of the ureter or a lesion of that organ could suspend the true secreting function of the corresponding kidney, while it left its percolating function intact, or even if any theoretical explanation of such a result could be given, the case would be quite clear, since the opposite kidney would have double secretive work to do, and the urine passed by the urethra would be scanty, with excess of lithates," &c. I have thus ventured to make a suggestion in reference to a point in connection with these cases which has hitherto failed to receive a satisfactory explanation. Pathological investigation has worked out all the facts connected with this lesion. I have merely speculated as to how these facts may be applied.

The possibility of a rupture of the ureter must not be lost sight of in association with injuries and wounds of the lumbar region.

Laceration of the kidney is a lesion more frequently recovered from than any corresponding one of an internal organ, and my connection with the hospitals of this town has been the means of furnishing me with several examples of this truth. Usually such laceration is caused by the application of direct force to the loins, as a crush between waggons, a heavy weight falling on the back, or a fall from a height, the region of the loins coming in contact with some prominent object. I will take a case which occurred at the Northern Hospital, where, though there were no means of verifying, by inspection, the nature of the internal damage, no reasonable doubt could be entertained as to what had occurred.

A dock labourer, aged 42, was in the summer of 1865 admitted into the hospital under my care for an injury to his back, caused by falling down the hold of a steamship on to the edge of a case of goods.

The patient was much collapsed, and there was a considerable contusion in the right lumbar region, without any breach of surface.

On partially recovering from his collapse, in the course of a few hours, he passed urine deeply discoloured by blood and small clots. The patient's condition gradually improved, though the urine showed traces of blood for nearly three weeks. On the day following the injury the urine contained some long worm-like casts which were clots that had been moulded within the ureter. These were not present after the third day. In addition, by the microscope blood-casts of the uriniferous tubes were occasionally seen. The treatment consisted chiefly in fomenting the injured part, as being the most comfortable application, and restraining any tendency to excessive hæmorrhage from the damaged kidney by the internal administration of the infusion of matico, a drug which, under similar circumstances, I have often found exceedingly useful. There was no indication that the urine was being rendered putrescent by blood-clots collecting in the bladder, and therefore no necessity arose to adopt irrigation. The diet consisted chiefly of milk and other bland nourishment, administered until all signs of hæmorrhage had ceased.

In reference to the collapse from which the patient

suffered on his admission, though it was extreme, I did not sanction the use of stimulants. In all injuries involving a laceration of an internal organ I recognise in the collapse which immediately attends them a most important provision against internal hæmorrhage. I am sure that many lives are sacrificed, under these circumstances, by injudicious friends plying the unfortunate sufferers with stimulants, forgetting that by thus arousing the heart's action to its normal force the risk is incurred of interfering with the process of clotting by which the vessels are sealed and excretion is immediately suspended. The latter is an important consideration, for if the laceration were sufficient to permit of urine escaping into the tissues about the kidney, this would be a fruitful cause of irritation, such as is seen when urine is extravasated amongst damaged structures. For some time afterwards the injured kidney must be little else than a percolator of water, minus the urinary salts, the excretion of the latter being provided for by a compensating action on the part of the opposite organ. It is to the immediate plugging of the renal bloodvessels, coupled with the fact that, if time is allowed, the uninjured organ is capable of doing double duty, that so many recoveries take place after rupture of the kidney.

Hence I much prefer meeting the collapse which, to a greater or less degree, invariably attends this injury by the use of external warmth, in the shape of hot blankets and bottles, and sinapisms, which by determining blood to the skin and establishing a diaphoresis, favour those processes of recovery and repair to which reference has been made. At the Northern Hospital and Infirmary, where I have seen several instances of this injury, I am sure that recovery of several patients was largely due to the carrying out of this principle.

In wounds of the kidney, hæmaturia is generally admitted to be a constant symptom. In ten cases described by Dr. Gustav Simon* it invariably occurred, and in several in very considerable quantities. A careful examination of the urine will frequently be found to afford valuable evidence as to the source of the hæmorrhage. This point I have referred to elsewhere in drawing attention to the observation of Mr. Hilton, that all clots in the urine should be floated out in water, with the view of ascertaining whether their shape in any way indicates their source; and, with the same object, urine containing blood should be invariably submitted to microscopical examination, as the discovery of blood-casts is of much importance in establishing a diagnosis.

The kidney is sometimes wounded by the discharge of firearms and other explosives. The surgical records of the Civil War in America, to which reference has been made, furnish numerous and authentic data upon which conclusions as to treatment may be founded.† An analysis of these data affords no instance where nephrotomy or extirpation of the wounded organ would appear to have been desirable. The reporter remarks that the practice which Larrey and Dupuytren recommended, of enlarging the lumbar orifices of wounds of

* *Die Chirurgie der Nieren*, 1876.
† *Medical and Surgical History of the War of the Rebellion*, Part ii., p. 159.

the kidney to prevent extravasation of urine internally, does not appear to have been followed, incision being reserved for the subsequent phlegmonous swellings which frequently formed in the loins.

The application of the generally admitted principles of treating wounds involving internal organs, appears, by the records referred to, to have been followed by a very considerable and encouraging success.

Some rare cases of hernia of the kidney through a wound in the abdominal walls are recorded; of this accident I have never seen an example. I quote the following from Pilcher,* with its reference:—

"June 3, 1873.—S. P., aged 25, was stabbed with a knife, in the left hypochondrium; two or three hours after a cough set in, which caused the kidney to protrude through the wound. At the end of twenty-four hours he presented himself at the clinic of Professor Brandt, in Klausenberg, having a pulse of 80, a temperature nearly normal, and being able to walk to a gallery to be photographed. On the fourth day after being wounded the kidney was drawn out and severed, after its pedicle had been ligatured. Rapid recovery resulted. At no time did he show symptoms of uræmia or peritonitis. The quantity of urine secreted increased daily while he was under observation. June 23rd he left the hospital, able to work as before." †

* *Surgical Operations upon the Kidneys*, by L. S. Pilcher, U.S.A.
† *Wiener Med. Wochenschrift*, 1873.

TWENTY-SIXTH LECTURE.

SURGERY OF THE KIDNEY.

AT the present day any treatise on the surgery of the urinary organs would be singularly incomplete if it failed to notice what had been done for the relief of certain affections of the kidney and the ureter which have proved to be beyond the aid of medical treatment. I therefore propose to bring briefly under consideration the somewhat scattered experience we possess of the operations which have been practised on this portion of the urinary track. Cases of this kind are exceptional, even with those whose opportunities of seeing this class of affections are abundant; recent improvements in operative surgery have undoubtedly added to the reasons rendering such proceedings justifiable.

So far as I can gather, operations on the kidneys have been practised under the following circumstances :—

First : Exploratory operations having for their object the actual inspection and manipulation of the suspected organ or its ureter.

Second : The puncturing or incising of the kidney for the evacuation of matter or fluid, and the removal of calculi.

Third : The complete ablation or removal of one of the kidneys.

First. Exploratory operations, by means of which the kidney has been exposed and examined, have on several occasions been practised; and, as illustrating this class of cases, I will refer to a record by Mr. Annandale, where aggravated symptoms of renal calculus appear to have been cured in this way. This case shows the difficulty there is in precisely determining the cause of renal irritation. Had the operation failed to cure, it would still have been of value in eliminating, at no great risk, an element of doubt which, so long as it existed, was an obstacle to successful treatment.*

The case was that of a woman, aged 36, in whom there was a strong suspicion of impacted renal calculus as being the source of the irritation from which she was suffering. As the patient's health was failing, and other means of relief proved futile, the kidney was cut down upon, by an incision along the outer border of the erector spinæ muscle, and examined, together with the greater part of the corresponding ureter. The operation was conducted with all antiseptic precautions. Though no abnormal condition of the organ was discovered or calculus felt, the relief obtained was complete, and justified Mr. Annandale's observation, that the case "proves that the kidney and ureter can be freely exposed and handled without serious results following the operation."

Somewhat similar to Mr. Annandale's case are two others which are referred to in a recent paper on this

---

* "Case of Aggravated Symptoms of Renal Calculus cured by an Exploratory Incision," by T. Annandale, F.R.S.E.—*British Medical Journal*, Dec. 19, 1874.

subject;* in each instance it is stated that no incision was made into the structure of the kidney. "Dr. Gunn, of Chicago, cut down upon and exposed the kidney in a male supposed to be suffering from renal calculus. No calculus could be felt when the fingers grasped the kidney; the kidney was not incised nor interfered with; patient recovered well from the operation." †

"In 1870, in the case of a woman who presented the ordinary symptoms of renal calculus, Mr. Durham, at Guy's Hospital, cut down upon and exposed the right kidney, but being unable to find anything abnormal about it, desisted from further interference."‡

Here, then, we have three illustrations of an exploratory operation on the kidney having been practised without any ill effects following.

Lastly, under this heading, we must include the puncturing of the kidney or its ureter by means of an aspirator or other fine needle, for the purpose of directly ascertaining the presence of a stone in either of these positions. This may be practised with impunity, and has been done on more than one occasion. I have elsewhere referred to Mr. Barker's case as being, I believe, the first recorded where this additional means of establishing the diagnosis of renal calculus was adopted. §

Second: The puncturing or incising of the kidney, for the evacuation of matter and the removal of

---

* "Surgical Operations on the Kidney," by Dr. L. S. Pilcher, Brooklyn, New York.
† *New York Medical Journal*, 1870, vol. xii.
‡ Pilcher, *Op. cit.*
§ *Vide* p. 266.

calculi. Examples of operations on the kidney undertaken with these objects will be found more numerous. The earliest case is quoted as that of the French archer, who, being under sentence of death, and suffering from a nephritic trouble, was delivered up to the Faculty of Paris for experiment. The kidney was cut down upon and a stone extracted. The patient recovered and lived many years afterwards in good health.

Amongst more recent cases of this kind I will refer to one recorded by Mr. Chauncy Puzey, where I had the advantage of witnessing the operation.[*] This patient, after a variety of urinary troubles, chiefly involving the urethra, showed signs of renal abscess, with indications that pointed to the soundness of the opposite organ. Accordingly, the kidney was cut down upon in a manner very similar to that adopted for opening the colon, and freely incised, a large quantity of matter escaping. I will conclude my reference to this interesting case by a brief extract from Mr. Puzey's remarks as bearing upon certain details in the performance of the operation: "Although the history of this case terminates with the death of the patient, I think I may claim that the operation was not in any way the cause of death; but that, on the other hand, it prolonged life, gave great relief, and would most probably have resulted in a complete cure had it not been for the diseased condition of the liver. Whether it would have been better to attempt complete removal of the kidney is a matter on which there

[*] *The Lancet*, Feb. 7, 1880.

may be some difference of opinion. In this case my own opinion is that it would have been impossible without tearing away part of the peritoneum. In a similar case I should certainly adopt the same plan as I did in this instance; but I should not take any trouble to re-establish the flow of urine towards the bladder, but rather trust to free opening and drainage, and hope for contraction of the kidney. When this case was read before the Liverpool Medical Society, a member suggested that with this end in view the organ might be freely cauterised and destroyed. This plan would, I think, be dangerous, but I see no reason why, as long as a sinus remained, the wound should not be from time to time laid open deeply and the exposed surface of the kidney well cauterised, thus doing just as much as, and no more than, would probably suffice for the desired purpose."

In the discussion which followed the narration of this case at the Medical Society, some speakers seemed to think that so long as any of the true secreting structure of the kidney remained it would be impossible for occlusion of the sinus to take place. In reference to the steps of the operation, Mr. Puzey concluded, "that the incision as for colotomy was easier, and exposed the kidney more freely. There must also be less division of nerves of the lumbar plexus; although in my case I presume that I divided the ilio-inguinal nerve, as the patient for a long time afterwards felt the right groin and scrotum quite numb."

A case of pyelitis, where incision into the kidney was successfully practised, is recorded by Dr. Habershon, which, with the remarks that followed, have an important bearing upon the point now under consideration.*

A proceeding of a somewhat similar nature to that resorted to by Mr. Puzey, but resulting in the extraction of a calculus, is published conjointly by Dr. Andrew and the late Mr. Callender.† The patient, a woman aged 44 years, was admitted into St. Bartholomew's Hospital with undoubted evidence of a renal calculus, which, by reason of the suppuration and irritation it was keeping up, was rapidly wearing her out. Mr. Callender made an incision into the right loin as for colotomy, and on reaching the kidney incised it vertically. The calculus which was discovered and removed consisted of lithates coated with phosphates, and weighed over an ounce, excluding the fragments which had been detached.

Commenting upon this case, Mr. Callender remarks: "This woman was made no way worse by the operation. She was failing before it was practised, and the end came as it probably would have come had no operation been interposed. Her case proved the facility with which the removal of a renal calculus may be effected; but in this instance the wasted muscles and the well-defined tumour were conditions which favoured the steps of the operation, and these could not, of course, be always reckoned upon. If I have again to practise it, there is no modification I would make (save under exceptional conditions) in the line of inci-

---

* *Clinical Society*, Jan. 23, 1880.
† *St. Bartholomew's Hospital Reports*, vol. ix.

sion, which should be planned to strike somewhat obliquely across the outer border of the quadratus. That border, when exposed, should be the guide for the deeper incisions towards the coverings of the kidney, which in these cases may be expected to have displaced forward the colon, the only structure which can have to be considered in cutting from the quadratus muscle. Having exposed the kidney, it may be conveniently sounded with a fine trocar and cannula; and if a calculus is present, the operator should be prepared, after exposing it, to detach it from some at least of its prolongations into the calyces, which prolongations may be subsequently removed; and this should be done before any but the gentlest attempts are made to remove the stone, for it is generally, as we know from post-mortem experience, locked fast in the pelvis by its branching portions, and cannot be extracted until, as has been said, some of them are broken off." In the performance of Mr. Callender's operation, antiseptics do not appear to have been employed; it is stated that there was no hæmorrhage, and the wound was left open.

More recently an instance has been recorded where a stone, weighing 1·3 ounces, was removed by abdominal section. Though it terminated fatally on the thirty-first day, there is much in the case both of interest and encouragement. As a preliminary to the operation, adhesion of the kidney-cyst to the abdominal wall was induced by the method advocated by the late Dr. Simon, of Heidelberg. The case indicates the necessity, at the time of operating, of being provided with cutting forceps sufficiently powerful to break up the stone, so as to permit, as Mr. Callender pointed out in his instance, of its being removed with the least possible damage to what remains of the kidney. Mr. May concludes his very interesting case with the following passage :—

"The advantages claimed for operations through the abdominal wall are the comparative thinness of the structures covering the tumour, and also that in this situation the incision being carried directly into the dilated pelvis, the substance of the kidney suffers less injury, and the risk of hæmorrhage is thereby greatly lessened. In the case of a large-branched calculus, it may be added that the attempt to withdraw it is made at much greater advantage through the pelvis of the kidney than in the direction of its branches through the substance of the kidney itself. The operation practised in this case was designed and carried out by Simon for cases of pyonephrosis and hydronephrosis, and has been more recently performed by another German surgeon, and the success attending it in their hands led to its adoption in this instance."*

I might refer to other illustrations of this proceeding (which has been named litho-nephrotomy), but these are sufficient for my purpose, and I will pass on to notice, in the third place, the complete ablation or removal of one of these organs.

Such an operation can only be entertained where there are substantial grounds for believing that the opposite organ is unaffected and is capable of under-

* *The Lancet*, July 3, 1880.

going those hypertrophic changes necessary to render it capable of performing the whole urinary secretion. That such a condition is possible, nay, frequently exists, is shown by experiment upon animals and pathological observation. No better illustration can be given of the marvellous natural powers we possess of adaptation for the preservation of life than is found in those instances where, one kidney having become irreparably damaged, the other, by an increase in its size, supplies the excreting surface that is required for the performance of double work, as its fellow gradually passes into a condition of useless but harmless atrophy. Surgery should decline to interfere in such a beneficent process.

Unfortunately, however, certain forms of kidney disease are attended by profuse suppurative changes. A state of fever is produced which surely and speedily, unless the cause is removed, undermines the constitutional powers of the patient and destroys life. In the suppurating kidney we have an exact analogy to that which is observed in the similarly affected joint and its attendant hectic. Surgery proposes to do for the suppurating kidney what is done for the suppurating joint—namely, to remove it. Let me proceed to notice under what circumstances the operation has been practised.

To the late Professor Simon, of Heidelberg, we are indebted for having first demonstrated that extirpation of a human kidney may be successfully practised. The operation was performed by lumbar section in 1869, for the relief of a urinary fistula; the patient was reported as being in good health seven years after. Since the publication of this case the operation has been resorted to on numerous occasions and for a variety of purposes; amongst the latter are included cases of malignant disease of the kidney, nephro-lithiasis, tubercular disease, cystic disease, for the removal of a floating kidney accompanied with constant distress, fistulous communications of the kidney with other organs, and lastly, for wounds of the abdominal walls, attended with hernia of the kidney.

From a careful examination of the cases in which extirpation of the kidney has been practised, it is at once obvious that it is an operation which can only be resorted to under very exceptional circumstances. When undertaken for malignant disease, we must be fortified by almost a certainty that the disease is confined to the one kidney it is proposed to extirpate. In cases of disorganisation of the organ, such as we see in suppurative tubercular disease, I think that incision through the loins, in accordance with the plan to which I have already referred, accompanied by thorough drainage, will generally be found the more suitable. For the removal of stones impacted in the kidney, it does not appear to me that ablation of the organ will afford greater advantages than making an opening with the knife sufficiently large to permit of the extraction of the calculus, should this be found practicable.

Though by the employment of antiseptics much can now be undertaken with comparative safety which a few years ago would have been considered inadmissible, we must not allow this consideration to carry us beyond those limits which a correct appreciation of pathology determines; and this rule necessarily imposes, as I have before pointed out, as a condition for undertaking the operation of extirpating a kidney, not only that the disease can be removed, but that the opposite organ is in a fit condition to carry on the whole process of urinary excretion. To say that nephrectomy is, under all circumstances, an unjustifiable operation, would be as improper as to assert that it is adapted to all the conditions to which it has hitherto been applied.

I will now pass on to notice the methods which have been employed for the removal of a kidney. A reference to the anatomy of the part concerned shows that extirpation may be accomplished in two ways : (1) by an incision through the loins; and (2) by abdominal section. The former plan has the advantage of allowing the organ to be reached from a point where it is uncovered by the peritoneum, whilst abdominal section not only gives greater freedom for the necessary manipulation, but permits of an examination of the diseased organ and its relations, which operators who have practised this method do not hesitate to acknowledge. The selection of the one or the other proceeding will depend upon the special circumstances of each case. In a case recorded by Mr. A. E. Barker,[*] where nephrectomy was employed for a movable kidney affected with encephaloid disease, ablation was readily accomplished by abdominal section. In a child, from whom Mr. Knowsley Thornton successfully removed the

[*] *Royal Medical and Chirurgical Society*, March 9, 1880.

kidney,* a similar mode of proceeding was employed; whilst in Mr. Lucas' case † of suppurating kidney, extirpation was effected by an incision through the loins.

Extirpation of the kidney by abdominal section appears to have been done by an incision, and with antiseptic precautions, very similar to those employed in ovariotomy; the organ being shelled out from behind the external layer of the ascending meso-colon. In Mr. Barker's case it is stated that "the ureter was tied separately, the pedicle in two portions by transfixion, with twisted silk." †

Dr. Thomas Savage, of Birmingham, records a case where nephrectomy was successfully performed by him for hydronephrosis. Abdominal section was practised; as the whole of the cyst could not be removed, a wire clamp was applied about the middle of it after its contents had been evacuated.‡

For the lumbar operation, by which the kidney is reached without opening the cavity of the peritoneum, the incision has been very similar to that adopted for lumbar colotomy, subject to such modification as the special nature of the case may require.

A good example of nephrectomy performed for a floating kidney giving rise to symptoms which were otherwise irremediable is recorded by Dr. A.W. Smyth.§

"The operation was commenced by making an incision in the right side of the lumbar region, extending externally from

* *The Lancet*, June 5, 1880.
† *British Medical Journal*, March 13, 1880.
‡ *The Lancet*, April 17, 1880.
§ *The New Orleans Medical and Surgical Journal*, August, 1879.

the crest of the ilium to the edge of the eleventh rib, two and a half inches by measurement from the median line of the spine and parallel with it. Internally the incision extended to the edge of the twelfth rib. The muscles and the transversalis fascia having been divided, search was made for the kidney, which was found in the umbilical region. The kidney, by pressure upon the abdomen, was forced into its place, and while held there by an assistant, the fascia covering the kidney was ruptured by the finger, and the organ was extracted without difficulty. While still in the wound, a strong ligature was passed round the renal vessels and other connections, at a distance of less than an inch—perhaps, about half an inch—from the hilus; and the organ was then detached. No elongation of the connexions of the kidneys was observable.

"Nothing worthy of special note—much less anything untoward—occurred during the operation. At its conclusion, two sutures were inserted, to bring the edges of the integuments together, in the upper part of the wound, the ligature being left hanging out of the lower part. The wound was dressed with a solution of carbolic acid, of the strength of one drachm to a pint of water. A hypodermic injection of half a grain of sulphate of morphia was administered, and repeated at bedtime.

"The kidney removed was found to be of normal size, but to be scarred with a deep cicatrix, extending, from the inferior and outer edge, obliquely up, and out, and apparently through the pelvis. The length of the cicatrix was about two inches and a half. It was evidently the result of the seton introduced (for the purpose of endeavouring to fix the kidney by adhesion), which had cut its way completely out of the organ.

"The operation has been followed by complete recovery; and the patient no longer complains of the trouble afflicting her, on account of which it was undertaken."

In a letter dated June 2, 1880, Dr. Smyth informs me that "the patient continues in good health. The operation can be performed with a facility which will surprise anybody that undertakes it."

For statistics relating to nephrectomy I must again quote from Mr. Barker's carefully drawn up paper, which I listened to with great interest, and the value of which I take this opportunity of acknowledging: "Of the twenty-eight attempted nephrectomies, six were done on a wrong diagnosis; in two for neoplasms the operation was left uncompleted. In the whole twenty-eight there were fourteen recoveries and fourteen deaths. Excluding the six for wrong diagnosis, there were thirteen successful and nine fatal; two of the latter were desperate, and left uncompleted."

I have thus endeavoured to take a brief glance at the circumstances under which the kidney has been subjected to operative interference.

TWENTY-SEVENTH LECTURE.

Tumours of the Bladder and Prostate — Scirrhus and Medullary Cancer — Non-malignant Tumours — Villous Growths — Palpation by the Rectum.

Clinically, tumours of the bladder and prostate, like those affecting other parts, are divisible into two groups—namely, simple and malignant. The former are capable of existing for a considerable period of time without grave ill-consequences, so long as by their presence they do not seriously interfere with the functions of the bladder; whilst the latter present all the well-known destructive features of cancerous growths.

First, in reference to cases and specimens illustrative of malignant tumours.

An interesting case of a rare disease—namely, scirrhus of the prostate—is recorded by Dr. Dickinson, from notes by Dr. Craigmile, as occurring at the Northern Hospital.*

"G. B., 47 years of age, a sailor, was admitted into the medical wards on Oct. 20th, 1876, suffering from chronic rheumatism. The pains in the joints soon passed off, but as he remained very weak, a more careful examination was made, and he then stated for the first time that he had pain and difficulty

* *The Lancet*, April 28, 1877.

in passing water. He had had gonorrhœa a year before, followed by stricture, for which he had been treated by instruments. The perinæum was hard and cartilaginous, and there were two fistulous openings there. The glands in both groins were considerably enlarged, especially on the left side, and all were of a stony hardness. On examination per rectum, a hardened mass was felt, corresponding in size and shape to an enlarged prostate, and so hard as at once to suggest scirrhus, especially when associated with such glands. No catheter could be introduced beyond the stricture, but as morphia suppositories were found to give him ease in making water and freedom from pain, no further attempt to cure the stricture was made. The other signs were those of persistent cystitis, and occasionally he passed blood. He got gradually weaker, and the cancerous cachexia became more marked. He died on the 12th January, 1877.

"The post-mortem appearances were the following:—The tissues at the base and sides of the bladder were all matted together and thickened. The prostate was about the size of a horse-chestnut, and when cut into had all the appearance of scirrhus. There were three glands lying along the right iliac vessels, much enlarged and hardened. The bladder showed well-marked signs of cystitis, both ureters were greatly dilated and thickened, and the kidneys were undergoing atrophy from the backward pressure of the urine; but all these changes seem to have been due to the stricture rather than to the disease of the prostate, since the prostatic portion of the urethra was of normal size, and the tumour did not seem to obstruct the outflow of urine. There was no appearance of cancer elsewhere, nor any other noteworthy change in any of the organs. Microscopic examination showed great dilatation of the tubes of the gland, with large collections of cells in them, as in ordinary glandular carcinoma, but there was exceedingly little infiltration of the muscular stroma, which seems to be characteristic; for Rindfleisch, quoting another authority, says it is confined to the glandular elements, and that the stroma remains passive. The enlarged glands were also cancerous when examined. The kidneys both showed well-marked interstitial nephritis."

I have introduced this case as illustrative of the features which led to the discovery of the disease during life, and its identification after death by microscopical examination. The patient presented the usual indications of scirrhus, as noted in other parts of the body—viz., extreme hardness of the prostate, enlargement of the adjacent glands, and the marked "cachexia" of cancerous disorders; in addition, we have, as incidental to all growths within the bladder, occasional hæmorrhage and signs of cystitis. I would observe that, in examining cases where there is a suspicion of a tumour growing within the bladder, it is important to notice the state of certain small glands and lymphatics which may be felt by the finger in the cellular tissue in front of the rectum. In malignant disease I have found them to be indurated, and thus obtained an important clue to the nature of the disease.

Passing to another variety of cancer—viz., the soft or encephaloid—I am able to give you a well-marked example which came under my observation, under circumstances where it was found necessary to relieve the most urgent symptom, and empty the bladder by means of the aspirator.

In January, 1874, I saw, with Dr. W. Little, a youth, aged 19, suffering from retention of urine. I was furnished with the following particulars:—For four weeks previously he had experienced difficulty in passing water, consequent on an attack of gonorrhœa, from which he was suffering early in December. Previous to the gonorrhœa he was in every respect in good health. On the 30th December the difficulty in urinating was so great that he applied to a surgeon, who relieved him by catheterism. From this date similar treatment under different hands had frequently to be resorted to.

On the 28th of January he came under the notice of Dr. Little, suffering from retention of urine; and, as the case presented certain peculiarities, I was also requested to see him. This I did on the following day. On examination, I found the bladder unusually prominent and largely distended. The perinæum was also much distended, and on introducing the finger into the rectum a similar condition was discovered. The prominent or bulging perinæum was unlike an abscess, as the swelling was elastic and of uniform consistence. The skin was neither inflamed nor discoloured, and there was no indication that extravasation had occurred. On introducing a full-sized catheter, no obstacle could be felt; the instrument appeared to take a natural course, but no urine escaped. On removing the catheter, the apertures were found blocked with a brain-like substance. As urine had not been passed for some hours, it was clear that the catheter had failed to reach the bladder. Retention being the prominent symptom, it was equally apparent that this must be relieved at once, and as the catheter had failed, it was agreed to puncture what appeared to be a largely distended bladder above the pubes. This was done with the aspirator, when about three pints of dark urine were removed, which afforded relief, and the abdominal prominence subsided. Not so, however, with the perinæum; this remained as bulging and as tense as before.

The state of the perinæum being evidently the cause of the retention, a free incision was made along the central raphé. This gave exit to a mass of brain-like matter mixed with clots in various stages of disintegration. The mass which escaped spontaneously, or was scooped out by the fingers, was sufficient to fill a pint vessel. The hæmorrhage was general and copious, rendering it necessary to plug the wound, and apply pressure by a T-bandage. By this treatment the urgent symptoms were relieved, and subsequently urine passed through the penis and the perinæal incision. The puncture made by the aspirator occasioned no inconvenience, and disappeared in the course of a few hours. On January 31st a large sloughy mass came away through the perinæal wound, and a smaller one on the 5th of February. On the 4th of February symptoms of exhaustion set in, and from that date the patient gradually sank, dying on the 8th of February.

A partial examination of the parts after death showed the bladder to contain a considerable quantity of the cerebriform-looking mass. It is to be regretted that a complete examination of the body was not made, as the precise origin of the growth was consequently not satisfactorily determined. The nature of the growth was undoubtedly encephaloid cancer, presenting the usual characteristics of this disease.

It will be observed that the aspirator at once relieved the retention of urine, and, further, it enabled us to arrive at a diagnosis as to the cause producing it, which at the first aspect was certainly obscure. The extremely rapid growth of the tumour is worthy of notice, the duration of the disorder, from the history of the case, occupying only a few weeks. Whether the growth originated in the prostate, or within the bladder, cannot, in the absence of a more complete examination, be definitely determined; but as the prominent symptom throughout was difficulty in passing urine, it may be inferred that the gland was the part primarily affected.

The class of cases—namely, malignant growths—of which the two I have recorded afford well-marked examples, unfortunately, is not within the range of cure, though much may be done to relieve the distressing symptoms which are usually met with; such means are, however, limited to mitigating pain, arresting hæmorrhage, and overcoming any obstacle to micturition which may arise. The application of these means has already been considered in connection with other conditions of the bladder attended with similar symptoms, though produced by other causes.

In the treatment of cancerous affections of the bladder, accompanied with hæmorrhage, Chian turpentine has recently been tried by Mr. Clay, of Birmingham, and with advantage in some instances he has reported.* I cannot say that as yet I have found any benefit from its use in this class of cases.

Amongst non-malignant tumours of the bladder are included those flocculent excrescences not unfrequently met with which have received the name of villous growths. Of these we have several specimens, one of which is represented in the accompanying drawing. (Plate H.) These growths, so long as they do not give rise to hæmorrhage, or obstruct the orifice of the urethra, may exist for a considerable time without causing much inconvenience. Occasionally, by the irritation they give rise to, they simulate stone. I saw a case some years ago where the weight of suspicion that there was a calculus was increased by a peculiar

* *The Lancet*, March 27, 1880.

PLATE H

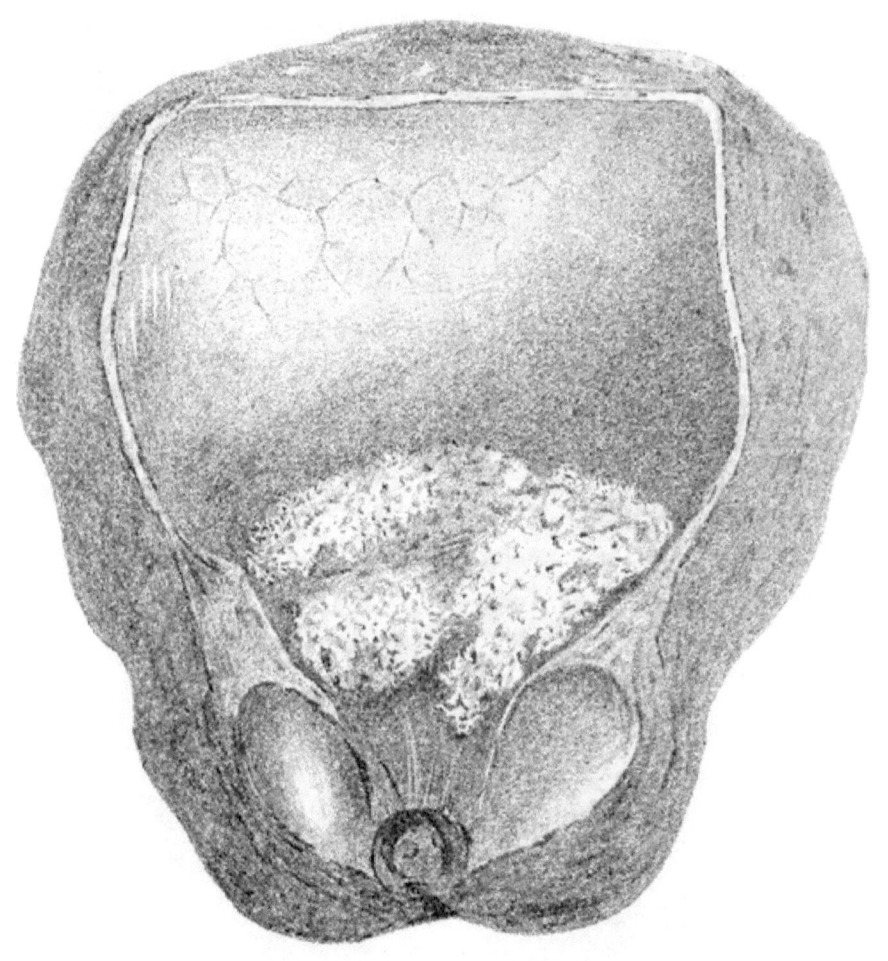

gritty sensation being felt as the sound was moved about the bladder; this, I believe, was explainable by the villi being encrusted with phosphates, as I have endeavoured to show in the Plate, where a similar condition is depicted.

As these growths are of so flocculent a nature, their presence during life cannot with certainty be ascertained, either by digital examination or by the introduction of the sound; the persistence of blood in the urine, aggravated by the passage of instruments into the bladder, together with the evidence which "the process of exclusion" affords, will point to the probable cause. Unmistakable proof, however, is not unfrequently afforded by microscopical examination of the clots and broken tissue which either escape spontaneously, or are entangled in the eye of a catheter that may have been introduced. Such occurred in the specimen (Plate H) removed from a patient of the late Mr. Long's, where, during life, the nature of the growth was in this way discovered. Again, in a case of Dr. Davidson's, where the patient was admitted into the Northern Hospital for severe hæmaturia, the existence of a villous growth, as the probable cause, was determined in a similar manner.

In an interesting paper on this disease,* Dr. Hudson draws the following conclusions:—

"1. Villous disease of the bladder is not so rare as is generally supposed, many so-called cases of chronic cystitis being probably due to it.

"2. Its diagnosis is most difficult, and can only

* *Dublin Journal of Medical Science*, June, 1879.

be arrived at after long observation and by a process of exclusion.

"3. Urinary deposits containing so-called cancer cells are very misleading, but the microscope is most valuable in detecting small portions of genuine villous growth.

"4. There should be no difficulty in detecting the growths in the female, as the whole internal surface of the female bladder can be readily explored with the finger after rapid dilatation of the urethra, when under the influence of an anæsthetic.

"5. Astringent injections are likely to be of use in the early stages, and before the growths have become pedunculated.

"6. The surgeon, while unsparing in the use of sedatives to relieve pain and spasm, should bear in mind the possibility of permanent cure by removal of the growth.

"7. Statistics show that the operation is neither difficult nor dangerous in the female; and there are good grounds for believing that when preceded by cystotomy in the adult male, it will prove justifiable and satisfactory."

In the treatment of villous growths, it must be remembered that when they prove fatal, it is by the severe hæmorrhage they occasion; consequently it is desirable to avoid lacerating them by any unnecessary introduction of instruments into the bladder. It must not be lost sight of that these excrescences belong to a class of tumours which in their nature do not offer any obstacle to removal; they do not show any tendency to ulcerate, to involve structures other than the mucous membrane, to implicate glands, or to invade the system generally. Where they prove fatal, it is by hæmorrhage, which might, there is reason to believe, be as capable of restraint, by mechanical means, as that which an open nævus would give rise to. Cases are recorded where these growths have been successfully removed.

Mr. Erichsen* alludes to an instance where Billroth successfully removed such a tumour from the bladder of a boy, by performing supra-pubic and median cystotomy at the same time, and, with the finger inserted at the supra-pubic opening, guiding a knife passed in through the perinæal wound.

Dr. Alexander, Surgeon to the Parish Infirmary, has kindly favoured me with the notes of a case in which he also operated successfully for the removal of a similar growth.

The patient was a female, 35 years of age, who appears to have suffered from a growth within the bladder for several years, the symptoms of which were aggravated from time to time by attacks of cystitis.

By the introduction of the finger into the bladder, Dr. Alexander was enabled to feel the villous excrescences, some of which were coated with phosphatic deposit. The patient being placed under chloroform, and the urethra dilated by means of an anal speculum, the wire of an écraseur was passed round the base of the largest growth, which, in this way, was removed. Some smaller excrescences were also detached by means of the finger, after which the bladder was washed out with a weak

* *Science and Art of Surgery,* 7th edition, vol. ii., p. 819.

solution of perchloride of iron. There was some hæmorrhage during the operation, and the urine contained a good deal of blood for the first week. The patient suffered but little constitutional disturbance, and incontinence of urine only remained for two days. After an interval, during which she remained in good health, a recurrence of her symptoms took place, when it was found that further growths existed; these were removed by the finger fifteen months after the first operation, since which date (October 5th, 1877) she has remained free from all symptoms of urinary disorder.

The appearance presented by these growths, as well as their microscopic characters, corresponded with those of the villous tumours to which allusion has already been made.

In addition to the cases I have mentioned, Professor Humphry[*] and Mr. Norton[†] have each furnished instances of the successful removal from the bladder of tumours such as I am now describing; I allude to them here as I consider that operative surgery is likely to do as much for the removal of innocent tumours of the bladder as has been done in the treatment of stone. The microscopical appearance of villous growths is well shown in the *Transactions of the Pathological Society*.[‡]

I recently saw, with Dr. John Bligh, of this city, a lady whose bladder was entirely filled with a villous growth, the removal of which was impracticable. Considerable relief followed the induction of a state of incontinence to urine and sanguineous discharge, which followed the instrumental dilatation of the urethra

---
[*] *Medico-Chirurgical Transactions*, 1879.
[†] *Clinical Society's Transactions*, vol. xii.
[‡] Vol. xxi., p. 239.

necessary to permit of the digital examination of the tumour.

Tumours presenting some of the appearances of these villous growths, but of a malignant nature, are occasionally met with. A specimen illustrative of this was recently exhibited at the Medical Institution, by Dr. Cameron. The patient was admitted into the Royal Southern Hospital in a very exhausted condition, suffering from hæmaturia, which had existed over three weeks. On examining the bladder after death. numerous excrescences, very similar to the true villous growth, were seen about the floor of the bladder. The portion of the viscus from which these sprang was thickened and indurated, the induration extending to the adjoining portion of the rectum. Several glandular enlargements were found between the bladder and the bowel.

Cases such as these have probably led some authors to regard villous growths as malignant; by their induration, their tendency to spread and involve other organs, such as the rectum, and to implicate glands. they present features which the true villous growth is not found to possess. Still, their occasional resemblance to them in one particular, and that, perhaps, the most striking, must not be passed by unnoticed.

It should not be forgotten, in speaking of the diagnosis of tumours of the bladder, that it is possible to introduce the hand into the rectum, and so to explore the contents of the pelvis. I have never had occasion to adopt this procedure, and for a description of the parts, as felt in this way, I will refer to the article on Palpation by the Rectum, by Mr. Walsham.* "Through the anterior wall [of the rectum] the hand first recognises the prostate, which feels like a moderately large chestnut. Immediately behind the prostate, the vesiculæ seminales may be distinguished as two softish masses, situated one on either side of the middle line. Internal to them the whipcord-like feel of the vasa deferentia can be readily traced over the bladder to the sides of the pelvis. The bladder is easily recognised, when moderately distended, as a soft fluctuating tumour behind the prostate; when empty, it cannot be distinguished from the intestines, which then descend between the rectum and the pubis. The arch of the pubis can well be defined when the bladder is empty."

This article further says:—"Dr. G. Simon, in a paper in the *Archiv für Klinische Chirurgie*,† states that repeated dilatation of the anus to its maximum does not destroy its contractile power, and that in no instance has permanent incontinence of fæces been the result."

This mode of examination having been recently introduced into practice, I have thought it necessary to bring it under your notice, as bearing upon the subject of the digital examination of the bladder, though at present I have no experience of my own to offer in reference to it.

* *Landmarks, Medical and Surgical*, by Luther Holden, p. 70, 2nd edition.
† Vol. xv., p. 1. 1872.

TWENTY-EIGHTH LECTURE.

Ulcerations of the Bladder — Tubercular and Cancerous Ulcerations — Perforating Ulcerations — Colotomy — Sloughing of the Bladder.

Ulcerations involving the bladder primarily, or by an extension of disease from surrounding organs, though not of frequent occurrence, will be most conveniently considered under a separate heading, as they require treatment very different from that appropriate to the various growths from which they may originate.

I will group them into two classes — namely, (1) surface ulcerations, where the lesion does not extend beyond the structures proper to the bladder; and (2) perforating ulcers, where a communication is established between the rectum or other of the pelvic viscera.

In the first group we have traumatic, tubercular, and malignant ulcers. In the second, malignant and non-malignant perforations.

Traumatic ulcerations may be caused by catheters retained in the bladder for retention of urine, and by the passage of foreign bodies, such as pieces of bone, which have been known to make their way through the coats of the bladder. Some years ago I made a post-mortem examination in the case of a sailor, admitted into the Northern Hospital, who had worn a silver catheter in his bladder for ten days previously. The instrument was a small one, and had been introduced with difficulty by the captain of his ship. The weather was exceedingly rough, and the patient suffered considerably, though he dare not remove the catheter, for fear of it being found impossible to re-introduce it. He died shortly after his admission. The urine passed during the short time he was in the hospital was loaded with blood. At the autopsy a deep ulcer, corresponding with the end of the instrument, was found in the roof of the bladder, the cause, no doubt, of the hæmorrhage which led to his death.

In retaining and securing instruments in the bladder, we should remember that it is sufficient to have the opening in the catheter just within the orifice, not only to avoid the infliction of such an injury as I have illustrated, but also to minimise the risk of giving rise to cystitis.

I have already alluded to a case of my own, where a piece of bone, by ulcerating through the walls of the bladder, had formed the nucleus of a calculus.

A very unusual instance of the passage of a foreign body into the bladder from the intestines by ulceration is recorded by Mr. Alfred Roberts, of the Sydney Hospital, New South Wales.* Here the patient, aged 47 years, had swallowed a piece of slate pencil two and a quarter inches long, which was subsequently successfully removed from the bladder by lithotomy. Commenting on this remarkable case,

* *Medical Times and Gazette*, July 30, 1859.

the author says:—" I have left no stone unturned to elucidate the truth in this very interesting case, and can only state that, after much hesitation, I have arrived at the conclusion that the pencil was swallowed by mouth, and made its way by inflammation and ulceration into the bladder."

Such, then, are a few illustrations of traumatic ulceration of the bladder.

Of tubercular ulceration of the bladder we are seldom without an example; indeed, from the frequency with which such cases come under notice here, there can be no doubt that tubercular disease of the urinary apparatus is a very common affection. We see it under a variety of circumstances. Tubercular deposit in the kidney or the ureter, not to speak of the bladder, is often attended with a degree of vesical irritability which leads to the suspicion that the patient is suffering from stone. The diagnosis of the first appearance of tubercle is exceedingly difficult, and often has to be arrived at, as you have seen, by what we have called a process of exclusion. Early tubercular disease of the urinary organs is most insidious. Where I have suspected that vesical irritability was due to it, rather than to any of those other conditions which suggest to a patient the desirability of obtaining surgical advice, I have derived considerable assistance from thermometric observations. The primary forms of urinary tuberculosis are, in addition to other slight objective symptoms, almost invariably associated with evening rises in temperature, which, so far as my experience goes, is not explainable by any other disorder of these organs.

In the diagnosis of urinary phthisis, I believe the thermometer is of the greatest value.

Tubercular ulceration of the bladder exhibits symptoms resembling stone. There is frequent and painful micturition and the urine generally contains both blood and pus. A careful examination with the sound is sufficient to establish the diagnosis. Not only is no stone felt, but the irregular masses of tubercular deposit may be made out as the bulb of the instrument passes over them.

There is one condition of the strumous bladder which is perplexing, and arises from the fact that tubercular deposit sometimes leaves behind a cretaceous cicatrix. I knew a patient who had been frequently sounded for stone, and on one occasion was nearly operated on, in consequence of his having a chalky scar on the walls of his bladder. I do not think there ought to be much difficulty in determining this condition with the sound.

In the treatment of tubercular ulceration of the bladder, much may be done by cod-liver oil, steel, tonics, and milk diet. Where there is nocturnal irritability and distress, an opium suppository is of the greatest assistance, not only in diminishing vesical spasm, but in preventing one of its ill-consequences—namely, dilatation of the ureters by back pressure of the urine. In the absence of signs of tubercular disease of the kidney, where ulceration had resisted other methods of treatment, it would be proper to place the bladder at rest for some time by allowing the urine as it was excreted to escape by an artificial vent, similar to the incision employed in lithotomy, rather than to permit it to collect in a bladder where its retention is attended with the pain likely to be produced by the contact of urine with a raw or an ulcerating surface. Such a proceeding has been practised with success.

Like other malignant growths, those affecting the bladder may end in ulceration. When this stage is reached, the loss by hæmorrhage, added to the distress which such a condition invariably gives rise to, soon terminates in death. The nature of the ulceration can generally be determined by the presence of one or other of the signs of malignancy, in addition to those indicating ulceration, which have already been referred to in my remarks on malignant tumours of the bladder.

In the second group we have two varieties of perforating ulcers—namely, the non-malignant and the malignant. Amongst the former we include ulcers which, though not cancerous in their nature, are exceedingly destructive, often not revealing themselves until in their progress they have involved other organs or spaces. Of these, some are remediable, whilst others are beyond the reach of surgical aid, assuming we were able to diagnose the course they were taking.

Belonging to the latter, is the very interesting case recorded by Mr. Bartleet, of Birmingham, where a perforating ulcer of the bladder made its way into the ileum, and caused death, as it were accidentally, by setting up peritonitis.* The ulcer, whilst confined to the bladder, as Mr. Bartleet remarks, appears to have gone through all its stages without presenting any

* *The Lancet*, Feb. 5, 1876.

symptoms, and whilst the patient continued to follow his accustomed occupation. A sudden lifting movement, which occasioned acute pain, most probably broke down a recent adhesion between the bladder and the bowel, and led to the extravasation of urine which was the cause of death.

Amongst the remediable perforations we must include cases such as that narrated by Mr. Hakes, for which colotomy was performed with complete success. In addition to the valuable records of this case,* which, in some respects, may be regarded as unique, I had the advantage of watching, with much interest, the patient throughout.

Briefly to summarise: the man was admitted into the Infirmary in a very deplorable condition, consequent on an ulceration between the bladder and rectum, which had existed for twelve months previously.

To remedy this, Mr. Hakes performed colotomy in the left lumbar space. The patient made an excellent recovery, and, for about three years, he enjoyed good health, acting as an omnibus conductor, and suffering very little inconvenience from his artificial anus.

Later on he fell into ill-health, and died from uræmia, consequent on extensive degeneration of the kidneys. At a post-mortem examination it was found that not only had the ulcer in the bladder soundly healed, but that the portion of the bowel below the artificial opening had become completely atrophied, being represented by a fibro-areolar cord, in which no trace of a canal could be found.

Such is a brief outline of this very interesting case,

---

* *Liverpool Medical and Surgical Reports*, vol. iii.—*Liverpool and Manchester Medical and Surgical Reports*, 1875.

which shows how successfully colotomy may be resorted to in non-malignant perforations of the bowel. It illustrates what rest may do, not only for the relief, but for the cure, of disease.

Though colotomy cannot be performed for malignant perforations with the hope of obtaining a permanent cure, it is often to be recommended as a safe means of arresting pain and prolonging life. It proved so in the following case, which has recently been in my wards, and which some of you had an opportunity of watching.

J. R., aged 53, was admitted into No. 1 ward on October 9th, 1877, suffering from a recto-vesical fistula of a malignant nature. The disease appeared to have commenced in the rectum ten months previously. On his admission the patient was in a very miserable and reduced condition. Within the rectum was a scirrhous ulceration, which communicated with the bladder by means of an opening through which a large-sized bougie could be passed. On introducing a catheter into the bladder there was first an escape of most fœtid flatus, followed by urine containing fæces in considerable quantity. The patient was suffering very severe pain, much of which was due to the frequent distension of the bladder with flatus. Frequent washing out of the bladder and rectum, together with anodyne applications, failed to give any permanent relief.

After a consultation with my colleagues, I opened the colon in the left lumbar region, and established an artificial anus. The relief that followed the operation was most marked, all the more distressing symptoms at once disappearing. For some time the patient improved; the disease, however, being evidently extensive, death took place from exhaustion on Dec. 1, 1877.

At a post-mortem examination, extensive ulceration of the rectum and bladder was found, the communication between the two cavities admitting the passage of a finger. The disease was of a scirrhous nature, and had probably originated within the rectum.

The operation entirely fulfilled my expectations—that is to say, it prolonged life and mitigated pain. My only regret was that it had not been performed at an earlier date; much suffering would in this way have been prevented, and I have very little doubt that it would have resulted in the comfortable protraction of the patient's life. It is better to act as soon as the nature and probable direction of the disease are made out, than to wait until the bladder is opened into by ulceration.

In concluding my observations on ulcerations of the bladder, I will briefly allude to those destructive changes which are sometimes observed to follow diseases of the spinal cord, and injuries attended with paraplegia. I do not refer to the damage which is occasioned by retained ammoniacal urine, and the necessary introduction of catheters, which are incidental to all cases of retention; though these may in a measure contribute to bring about morbid changes in the coats of the bladder, yet they are not sufficient to explain that rapid disorganisation of the viscus which is occasionally seen under the circumstances I have indicated. Such a condition is illustrated in the following case, a patient under the care of Dr. Glynn, in the Royal Infirmary, whom

I saw for the purpose of making an examination of his bladder.

T. R., aged 21, a porter, had previous to his present illness enjoyed good health. There was no history of syphilis. Two days before his admission to the Infirmary, when at work, he was seized with pain in the bowels. He walked home, took a dose of castor-oil, and applied a mustard poultice to the abdomen. In the night he tried to get up, and found that one leg was useless and numb. In the morning both legs were numb and absolutely powerless, and he was unable to pass his water. When brought to the hospital it was found that there was complete loss of power and sensation in the lower extremities. The urine had to be drawn off, and was found to contain pus and mucus. It was alkaline, and in the course of two or three days large quantities of blood were found mixed with it. Extensive sloughing of the bladder followed, and for some time before death all the urine was passed by the rectum.

On post-mortem examination, the spinal cord in the lower dorsal region was found softened. On section the distinction between grey and white matter was ill-defined. Under the microscope the large cells in the grey matter were much altered in shape, and dilated vessels and leucocytes were observed in large numbers. The coats of the bladder had sloughed, and abscesses had formed around it, through one of which a communication was established with the rectum.

Cases similar to this, where sloughing of the bladder has taken place, will be found recorded by various authors.*

Such destructive changes as these are probably dependent, as Charcot suggests, upon irritation of

---

* *A Treatise on Injuries and Diseases of the Spine.* By R. A. Stafford. 1832.

certain parts of the spinal cord, and more particularly the grey matter.*

* "How are we to understand so rapid a development of the inflammatory lesions of the urinary passages after acute affections, spontaneous or traumatic, of the spinal cord? Manifestly, the paralytic retention of the urine cannot here be pleaded, at least, not as the sole, nor even as the predominant, pathological element. Neither is it possible to attach great weight to the opinion (*Traube, Munk,* '*Berliner klin. Wochensch.*,' p. 19, 1864) which would attribute the urine changes, in such circumstances, to the introduction of unclean catheters, carrying vibriones. In point of fact, the introduction of vibriones into the bladder could only be a chance occurrence, whilst the appearance of ammoniacal, sanguineous, and purulent urine, in the course of acute myelitis, is, like the production of eschars, what may be termed a regular fact."

"The notorious insufficiency of the pathogenic conditions just enumerated renders it at least highly probable that there is a direct action of the nervous system engaged in the production of the affection of the urinary passages which we are considering. The cause of the affection, as of the other trophic lesions which often show themselves at the same time, would therefore be the irritation of certain portions of the spinal centre, and more particularly, no doubt, of the grey substance."—*Diseases of the Nervous System.* By J. M. Charcot. New Sydenham Society. 1877.

TWENTY-NINTH LECTURE.

Circumcision—Deformities of the Frænum and of the Meatus—Hypospadias—Epispadias—Patent Urachus—Absence of Bladder—Amputation of the Penis.

I shall, to-day, speak of and demonstrate an operation which has its origin in divine ordination, having been practised both by priests and surgeons since the day it was enjoined upon mankind that the males should be circumcised.

How it came to pass that circumcision should be decreed, which at first sight would seem to suggest an imperfection in creation, is a direction of enquiry foreign to my purpose; suffice it to remark that the removal of a portion of the foreskin appears to be a wise precaution against changes in the part, the result either of errors in development or of disease, which might interfere with comfort or health. From a clinical point of view, for these reasons, circumcision is rendered a necessary or an expedient operation.

It is to be recommended in all cases where the orifice of the prepuce is so contracted as not to permit of the glans penis being uncovered; in some persons the degree of contraction is so great as barely to permit the introduction of a fine probe. Here micturition is as much interfered with as if there were a stricture at the meatus of the urethra. To remedy this it is the practice of some surgeons to slit up the prepuce on a director. I do not think this is a good plan, as angles are left which sometimes prove very inconvenient. On more than one occasion I have had to perform an imperfect circumcision to remove the annoyance caused by these angular projections of the slit-up foreskin.

In addition to cases such as these, where removal of the contracted foreskin is a necessity, the operation is expedient where the prepuce fits the glans with some tightness or is unnaturally long or superabundant. In the former instance, not only is great inconvenience experienced, but the individual is exposed to the risk of his glans becoming strangulated by the retracted foreskin; this is technically spoken of as a paraphimosis. In the latter instance, the superabundant prepuce, by retaining secretions, subjects the person to attacks of balanitis; further than this, a redundant foreskin furnishes a convenient receptacle for inoculable virus. In reference to this point, Mr. Jonathan Hutchinson has demonstrated that the circumcised Jew is less liable to contract syphilis than the uncircumcised Gentile. "No one," he remarks, "who is acquainted with the effect of circumcision in rendering the delicate mucous membrane of the glans hard and skin-like will be at a loss for the explanation of this circumstance."* And what is true of syphilis is equally true of other diseases capable of being spread by contagion or roused into activity by irritation. It is exceedingly rare to meet with cases of

* *Medical Times and Gazette*, Dec. 1, 1855.

cancer of the penis in persons who have always lived with the glans uncovered.

In addition to these local conditions, which render it desirable, circumcision is sometimes necessary for removing the cause of reflex irritation felt elsewhere. How frequently we find in children nocturnal incontinence of urine provoked by an elongated prepuce, and effectually cured by its removal. Professor Sayre has pointed out how such an abnormal condition of the foreskin is the cause of various reflex pains which have, in the absence of the proper explanation, been regarded as indicative of hip-joint disease. I have seen him demonstrate the truth of this, and cure the apparently anomalous pains by removing a contracted prepuce. Under these circumstances, then, the operation of circumcision is to be recommended.

It has been suggested, where the contraction is not great or very inconvenient, that the prepuce may be gradually stretched to suitable dimensions by directing the individual to employ retraction. This is not prudent advice, especially in the case of young persons, for reasons which are obvious.

Dilatation of the prepuce is occasionally recommended where the patient is highly scrofulous, or is a "bleeder," and it is desirable to avoid making an incision. Various instruments have been devised for this purpose; of these the one described by Mr. R. W. Parker will be found extremely convenient.*

It has fallen to my lot within a short time to treat two varieties of stricture of the urethra which have

* *British Medical Journal*, July 19, 1879

resulted from improperly performed circumcision, and as they indicate the necessity of certain precautions being taken to prevent such inconveniences following, I allude to them here.

In two instances, the extremity of the glans penis, including the meatus, was wounded in making the section of the prepuce. In one the division had been made without the use of forceps, or other similar contrivance to include the foreskin, and secure retraction of the glans behind the line of section, the operator contenting himself with merely holding the portion to be removed between the finger and thumb, and then making the amputation. This is obviously a hazardous mode of proceeding. In the other case, where the glans had been wounded, there appears to have been considerable œdema of the prepuce, and consequently difficulty, in its swollen condition, of determining the precise position of the glans within it. Amputation of the prepuce had been performed, and the end of the glans included. Under such circumstances it is better first to slit up the prepuce with a probe-pointed bistoury, and, after having exposed the glans, to complete the circumcision with a pair of scissors. I have frequently done this when, from chronic balanitis and adhesions, it has been found impossible to determine the relative position of the parts.*

---

* "In some phimoses, the prepuce becomes so thickened and at the same time so elongated, that it resembles the body of the penis, and has led some into the mistake of supposing they had cut off a portion of the penis itself, when it was only a monstrous phimosis."—*On the Operations of Surgery*, by Samuel Sharpe, 7th ed., 1758.

In both of these cases there was a cicatricial stricture of the meatus, which I treated successfully by division with a tenotomy knife.

The second variety of stricture to which I have referred is rare, as it appears to have escaped the notice of the majority of the writers of systematic treatises. It is caused by the prepuce being divided too high up, or what amounts to the same thing, being drawn down too much over the glans penis before being included in the forceps for the purpose of making the necessary section. On bringing together the parts with sutures, the tension on them is so great as to cause ulceration, and to leave behind a broad cicatrix capable of exercising a contractile pressure on the under surface of the urethra sufficient to impede micturition, and to cause other discomfort.* I had to resort to a plastic operation to remedy the inconvenience that was thus occasioned by a too freely performed operation.

In circumcising a patient, remember to break down all adhesions between the glans and the prepuce. They are generally caused by attacks of balanitis, and sometimes are so firm that force has to be exercised to effect a separation. This, however, must be done.

In all cases of suspected ulceration of the glans penis, where retraction cannot be practised, it is a safe proceeding either to circumcise or slit up the prepuce, for the purpose of ascertaining the nature and extent of

* "The cut cuticular surface contracts on healing, and if a little prepuce has been purposely left at the patient's request, the new orifice is perhaps so tight as to be unable to pass over the corona during erection."—*Diseases of the Genito-Urinary Organs*, Van Buren and Keyes, p. 11.

the diseased action. In three cases, two in hospital practice and one in private, the doing of this revealed a state of ulceration which was hardly expected. In two of these patients the urethra had been opened into and a considerable portion of the glans destroyed. In the third instance cancerous ulceration was discovered. The only objection that can be raised against the practice is, that in cases of syphilis the recently cut surfaces may become inoculated with the virus; this, however, is as nothing compared with the damage by ulceration which may be going on. So long as you have the sores open before you, their extent matters but little.

The state of the frænum also requires consideration. Where it is not too tight it is better not to divide it, as after circumcision it is generally found to adapt itself to the altered relation of the parts. When, in the flaccid condition of the penis, it is so tense as to be on the stretch, or to depress the extremity of the organ, it is, in the adult, always well to divide it, otherwise the patient incurs a risk of rupturing it. On dividing it, a small artery usually spouts freely, and requires either twisting or a ligature. I have known in an anæmic young man serious hæmorrhage from rupture of the frænum. In this case several hours elapsed between the accident and my seeing the patient, whom I found in a state of collapse, with his bed and clothing saturated with blood. He remained in an alarming condition for some days, but eventually made a good recovery.

It has been alleged—I do not know with what

proof—that a penis may be so deformed during erection by a tight frænum as to interfere with emission. I will relate the particulars of a case which appears to me to bear upon the question, and I will leave you to draw your own conclusions.

A gentleman consulted me under the following circumstances. Eighteen months previously he was circumcised for a congenital phimosis; the surgeon who operated told him that, though the frænum was tight, as he was about to be married this would probably come right.

When I saw him he had been married twelve months, during which period not only had his wife failed to conceive, but connection had been attended with considerable discomfort, owing to the tense condition of the frænum. Naturally enough, he disliked again resorting to surgical interference, and hence this delay. I found him with the shortest and densest frænum I had ever seen. It was only a wonder to me, having regard to his development, that a rupture had not taken place. I divided his frænum, and within twelve months I heard that he and his wife had attained the object of their ambition.

Just a word about division of the remains of the frænum when it is undermined by chancroid ulceration, and a bridge of tissue alone is left. To snip it across with a pair of scissors saves a few days of ulceration; but this slight operation necessitates some bleeding if the artery is not obliterated by adhesive inflammation. As this is a small operation, which often has to be done in the consulting room, it is desirable, to say the least, to avoid soiling the patient's linen with blood. This is easily avoided. Take a finely-pointed piece of wood, and dip it slightly in nitric acid, and just touch the frænum on its cuticular aspect, at the site of proposed division, until it is blanched all round; then snip it across with your scissors, which you can do without hæmorrhage. By this means the frænum is divided in the course of a minute or so, without bleeding or even pain.

There is a congenital condition of the meatus which, like a phimosis, is a not infrequent cause of incontinence of urine in children—namely, an unnatural smallness of the orifice of the urethra. I have seen not only micturition in this way impeded in a young child, but considerable irritability caused, both of which inconveniences have been removed by division of the meatus, as referred to in Dr. Farquharson's paper on incontinence of urine.*

In addition to these states of the penis, which require slight operations, you will be consulted for other congenital imperfections of these parts. These deformities most commonly involve the end of the urethra.

The terms hypospadias and epispadias respectively denote a congenital deficiency of the lower and upper walls of the urethra. The former variety is by far the more frequent. In the cases I have shown you the deformity was limited to the orifice of the urethra, there being an absence of the frænum and a slit-like opening to represent the meatus. Where the imperfection is so limited as this, I am not aware that it gives rise to any special incon-

* *The Practitioner*, July, 1879.

venience. I am tolerably sure, however, that such persons are especially prone to become inoculated with venereal virus, the slit-like patency of the orifice forming a ready receptacle for it. In private practice I have recently seen two cases of indurated chancre in the urethra of persons suffering from slight hypospadias. I do not think that this consideration alone would warrant us in endeavouring to rectify an arrangement of parts which is not attended with other inconvenience.

In a boy that was in No. 2 Ward, where the hypospadiac condition was somewhat different, I endeavoured to remedy it by an operation, and I believe with success. This was rather a case of atresia, or imperforation of the glans, the patency of the urethra being limited to a small opening in its lower wall one inch and a half from its proper termination. He came to me for a difficulty in urinating, and as I thought it possible, in addition to this, that the use of the penis as an effective genital organ might be interfered with, I considered it proper to endeavour to restore the continuation of the channel. I succeeded in doing this by introducing a probe through the small urethral opening beneath the penis, into a cul-de-sac of the canal, and then making a passage with a tenotomy knife as nearly as possible in the right direction through the glans penis. This was subsequently kept open by the frequent passage of bougies. The boy micturated through the artificial urethra I had thus made for him, and the original opening on the under surface of the penis got still smaller.

I was arranging to close this by a slight plastic operation, when the boy left the Infirmary, considering himself quite well. I have not seen him since, but I was so well satisfied with what had been done, that I should feel disposed to repeat the proceeding.

In epispadias we have a deficiency of the roof of the urethra. Where this is combined, as it too often is, with fission and extroversion of the bladder, no more deplorable condition can be imagined, for not only does the individual, more frequently of the male kind, possess all the desire of the sex which he is unable to gratify, but further, the function of micturition is carried on in such a way as to be a constant source of personal distress and annoyance to others.

Surgery has done something to mitigate this distressing condition. By a series of plastic operations, Mr. John Wood has succeeded, in a number of cases, in rendering the condition of such persons endurable. For further information as to the details of these operations I must refer you to Mr. Wood's paper.*

An important communication on the treatment of malformations of the bladder has also been recently made by Mr. J. Greig Smith, of the Bristol Royal Infirmary, which may be advantageously referred to.†

In addition to these attempts to remedy epispadias by plastic surgery, some practitioners have sought to deviate the flow of urine into another

---

* *Medico-Chirurgical Transactions*, vol. lii.

† " Two Cases of Successful Operation for Extrophy of the Bladder, by a New Method."—*British Medical Journal*, Feb. 7, 1880.

channel, such as the rectum, hoping thereby to improve matters. As bearing upon this mode of treating these deformities, I would mention a paper by Mr. Thomas Smith.*

Of mechanical appliances for the relief of the condition in question I can speak favourably of an apparatus made by Tiemann, of New York, which consists of a metallic or hard rubber shield for application over the extrophied bladder, to the lower extremity of which is attached an elastic tube, leading to a soft rubber pouch to collect the urine; this is buckled to the thigh or may be carried still further downwards.

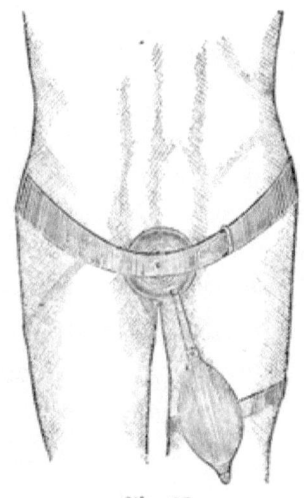

Fig. 36.

A case of congenital absence of the bladder is recorded by Mr. Vost, who appends the following note :—"A cavity to hold the urine could readily

* "An Account of an Unsuccessful Attempt to treat Extroversion of the Bladder by a New Operation."—*St. Bartholomew's Hospital Reports*, vol. xv.

be made by inverting a flap from the abdominal wall; but the absence of a sphincter would make the receptacle useless. All attempt at operative procedure was therefore abandoned."*

The only other abnormal condition to which I shall refer is one not often met with—namely, a patent state of the urachus. This structure extends from the apex of the bladder to the umbilicus, and retains the tubular character of the allantois till about the thirtieth week of fœtal life. Subsequently it becomes obliterated, and ceases at birth, except in some few instances, to have any tubular connection with the bladder. Where the patency of the tube remains after birth, the patient is subject to the discharge of urine from the navel, and a urinary fistula remains.

To remedy this defect, various means, similar to those employed in the treatment of other fistulæ, are recommended, including the application of cauteries and plastic operations. In reference to the treatment of this condition, Dr. Charles has urged the importance of removing any source of impediment in the urethra which may have been causing a back-flow of urine, such as a phimosis or a urethral calculus. Dr. Charles remarks that "to Professor Redfern is to be ascribed the credit of recommending circumcision; a novel plan for the cure of this abnormality, and one founded, as I have shown, on a consideration of its true nature." †

* *The Lancet*, August 14, 1875.

† "The Treatment of Patent Urachus."—*British Medical Journal*, Oct. 16, 1875.

An interesting note on a case of umbilical urinary fistula is recorded by Mr. F. Cadell, where, in a female child, a cure was attempted by establishing, temporarily, a state of vesical incontinence by dilatation of the urethra. This plan is sufficiently suggestive to require notice in connection with the treatment of this imperfection in development. *

As bearing upon the whole subject of the development of the urinary organs, and the errors proceeding from it, a recent article by Professor Watson, of Owens College, is worthy of careful study.†

Amputation of the penis is occasionally rendered necessary for malignant disease. The plan generally practised consists in sweeping off the organ at a point above the disease, and then allowing the stump to heal by granulation. To prevent retraction of the urethra, the mucous membrane is slightly slit up along its under surface, and the edges are then secured by sutures to the sides of the wound. This mode of proceeding is open to two objections: first, the orifice of the urethra is inconveniently situated when the amputation is close to the body, as urine, instead of being ejaculated, trickles down over the scrotum, and much distress by excoriation is thus occasioned; second, the cancerous organ is only incompletely removed—that is to say, more or less of the crura, probably infected, is left behind. To remedy this another proceeding has been adopted which I have

---

* *Edinburgh Medical Journal*, September, 1878.

† "The Homology of the Sexual Organs, illustrated by Comparative Anatomy and Pathology."—*Journal of Anatomy and Physiology*, October, 1879.

seen practised with much success. I am indebted to Dr. Gouley, of New York, for a description of this operation.

It consists in making an incision to embrace the root of the penis, and from the lower angle of the ellipse thus formed continuing the incision to half an inch in front of the anus, so as to bisect the scrotum. The spongy and cavernous bodies are separated and a urethra of two inches and a half is left. The crura are then dissected out from their attachments to the ischia, so that in this way the penis is completely removed by its roots. The urethra is next slit up and attached to the margins of the perinæal wound, horse-hair being recommended for this purpose. The remainder of the wound is closed by the ordinary interrupted suture. In the case from which this description is taken, on the completion of the operation a sort of vulva was formed, the patient eventually urinating with an excellent stream. The testes were left intact.

In his letter to me (dated October 4, 1879), Dr. Gouley goes on to remark, " I believe that in such cases (epithelioma) it is safer to remove the penis entire in this manner. So far as I know, Demarquay was the first to excise the penis, crura and all. It does not appear that Delpech, Lallemand, or Bouisson did more than cut the cavernous bodies close to the pubes."*

* " I find that in 1832, Delpech, of Montpellier, amputated the entire penis and bisected the scrotum, so that each testicle was enclosed by the aid of sutures in a separate scrotum, the extremity of the urethra being placed at the commissure of the bifid scrotum. The result was satisfactory; the patient was able to urinate well in a crouching position."—Lecture on Amputation of the Penis, by Dr. Gouley.—*Louisville Medical News, U.S.A.*, Sept. 15, 1877

There is another mode of amputating the penis which presents certain advantages. It was first suggested to me by Mr. Chauncy Puzey. It consists in performing a Cock's operation, so as to ensure the patient passing his urine through the perinæal wound. When this has been accomplished, and the stream of urine diverted, amputation of the penis may be performed as close to the body as possible, and the stump left to heal over by granulation. In this way the trickling of the urine from the cut extremity of the urethra over the scrotum is avoided, the patient micturating through the perinæum just as comfortably as he does after Cock's operation. It is necessary in these cases to instruct patients in regularly catheterising their perinæal opening so as to prevent any contraction taking place. I have seen this plan practised with complete success.

THIRTIETH LECTURE.

VARICOCELE — TREATMENT — PALLIATIVE AND RADICAL.

VARICOCELE is a morbid condition of the spermatic cord which so often incidentally comes under our notice in connection with diseases of the urinary system, that I purpose making a brief reference to it.

The term is applied to a dilated or varicose state of the veins of the cord, and consequently, like similar affections of these bloodvessels in other parts of the body, it varies much in degree and in the amount of inconvenience it causes.

Strangely enough, you will often notice that the symptoms attending varicocele are in no way proportionate to its extent; that is to say, you will find patients with a small varicocele complaining very much, whilst, on the other hand, some large ones appear to occasion little or no inconvenience. And this points to the relation in which the affection stands to the nervous system, in causing effects which are seen in the dismal and often absurd forebodings so frequently indulged in by the subjects of it. Hence, in treating it, or rather in managing the persons who have it, you will do well to remember Sir James Paget's observation, in his remarks on this affection, that it may be " a mental error, not a bodily one, that needs cure." *

* *Clinical Lectures and Essays.*

A varicocele is readily recognized. It usually commences shortly after puberty, and may be caused by any violent exertion, such as a strain. Some persons appear to have an hereditary predisposition to dilatation of the veins, and in these we have the disease appearing with apparently no other explanation to account for it. The left cord is more frequently affected than the right; this is probably due to the greater length of the veins on that side. To the feel the dilated veins have been aptly likened to a bag of worms. In extreme cases the veins almost conceal the testicle, but it is rare that they occasion any structural alteration in it. Like other varicose veins, they are influenced by the position of the body; when unsupported, and after much standing or walking, they cause a sensation of weight, if not of positive pain, to be felt. A varicocele is not likely to be mistaken for anything else. The only points in which it resembles hernia, so far as I have noticed, are, that it partially disappears when the patient lies down, and that something like an impulse on coughing may now and then be obtained.

The treatment of varicocele is palliative and radical. In the former, our object is, by artificial means, to provide a support for the distended veins, to prevent them enlarging or producing symptoms by their distension. In the latter, by setting up adhesive inflammation we endeavour to bring about their obliteration. These principles are similar to those we recognise in the treatment of the more familiar instance of varicose veins as seen in the legs.

It is the practice to apply support to the veins by a variety of expedients. Of these I may mention the late Mr. Wormald's plan, the description of which I will give in his own words, as being, from its simplicity, useful for reference. "A ring, about an inch in diameter, made of soft silver wire, of a suitable thickness, was padded and covered with washleather. Through this I drew the lower part of the scrotum, whilst the patient was in the recumbent position and the veins comparatively empty; then pressed the sides of the instrument towards each other with sufficient force to prevent the scrotum escaping. The use of this instrument every morning before the patient rose from his bed enabled this gentleman to walk nineteen miles on the third day after the first application; and although he has for six years worn an instrument of this description, he has never experienced the least inconvenience."[*]

Then we have the ordinary suspensory bandage, which in slight cases suffices. In more extreme varicoceles there are certain modifications of this appliance, which will be found useful.

Recently, Mr. Keetley has described a form of suspensory bandage, which is figured in the accompanying sketch.[†] (Fig. 37.) It is made in three sizes by Messrs. Arnold & Sons, of West Smithfield, and can be recommended.

A method of suspending the testicle and giving support to the enlarged veins of the cord was described

---

[*] *London Medical Gazette*, April 28, 1838.
[†] *The Lancet*, May 24, 1879.

by Mr. Morgan some years ago. As I have used this apparatus successfully in some very bad cases of vari-

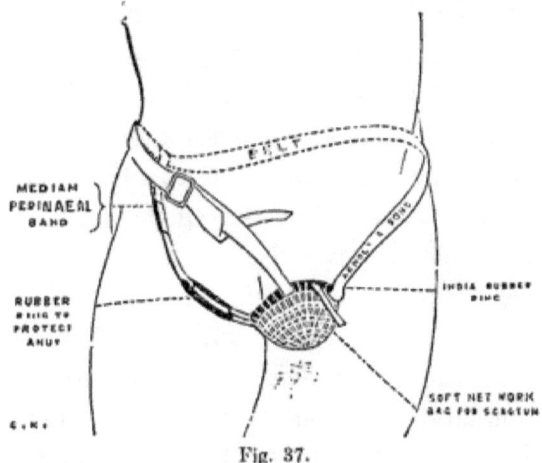

Fig. 37.

cocele, I will give Mr. Morgan's description of it as illustrated by the figure accompanying his paper.*

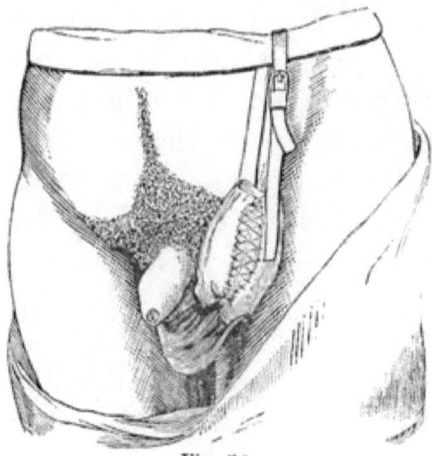

Fig. 38.

"The testis is shown in the suspender, which

* "On the Treatment and Cure of Varicocele by Suspension of the Testis."—*Dublin Quarterly Journal*, vol. xlviii.

consists of a piece of web, about 3½ inches wide at one end, 4½ inches long, 4 inches wide at the other, and cut gradually tapering to the narrower end. A piece of thick lead wire is stitched in the rim of the smaller end, and the sides are furnished with neat hooks, a lace and a good tongue of chamois leather, two tapes being sewn along the entire length of the web, which are afterwards attached to the suspending belt. The application is easily made by the patient in the morning before rising, and, when the parts are relaxed, laying the affected organ, while in the dependent position, in the suspender, and lacing up the hooks with a moderate degree of tightness, then raising it and attaching the tapes to the suspending belt previous to rising from bed. A certain amount of discretion must be used as to wearing the suspender for the first few days; it should not be kept on constantly; the parts should be sponged night and morning with cold water or a cold lotion, used so as to fortify the skin, as any chafing must be avoided. In all cases the suspender is best omitted at night. So great is the convenience afforded, that the gentleman from whose case the illustration is taken is now shooting in Scotland, able to enjoy himself and go through a day's hard walking without inconvenience."

Some practitioners prefer treating this affection by means of a truss, with a pad over the external abdominal ring. Care should be taken to select an instrument capable of giving sufficient support to the veins without exercising such an amount of pressure on the artery as to stop circulation.

By one or other of these appliances by far the greater number of varicoceles can be completely relieved.

I will now pass on to notice the circumstances under which it is justifiable to resort to operative measures, with the view of permanently obliterating the enlarged veins. Like all operations of this nature, a certain risk is incurred, which must not be too lightly estimated; and so long as some simple palliative measure is sufficient to render the patient comfortable, no other measures should be attempted. In a recent article Dr. Ogilvie Will[*] sums up the circumstances which would justify us in resorting to an operation as follows:—

1. If the varicocele be very large or increasing.
2. If the testicle be atrophied.
3. If acute pain be complained of.
4. If the patient be disqualified from entering the public service.
5. If the stability of the mental faculties be endangered.

On any of these grounds we are warranted in endeavouring to effect an obliteration of the dilated veins. Where the testicle shows any signs of commencing atrophy, Mr. Barwell[†] has shown that an improvement in its condition immediately follows the successful application of radical treatment. My own experience is corroborative of this.

Various means have been resorted to for effecting

---

[*] "The Influence of Varicocele on the Nutrition of the Testicle."—*The Lancet*, May 15, 1880.

[†] *The Lancet*, June 12, 1875.

obliteration of the veins, including the subcutaneous ligature, as practised by Mr. Lee, the introduction of needles and compression with twisted thread, and subcutaneous division with the knife. These proceedings are all either tolerably familiar or are fully treated of in the text-books. In a recent communication[*] Mr. A. Pearce Gould advocates the employment of the galvanic écraseur for the obliteration of the veins, and illustrates its feasibility by several successful cases. I have not yet had an opportunity of testing this mode of operating, but it appears to me to possess certain advantages, both as regards safety and efficiency.

In conducting these operations care must be taken not to interfere with the vas deferens or the spermatic artery; the former, on account of its whipcord-like feel, is very readily kept out of harm's way.

In concluding my remarks, I will mention a mode of operating which on two occasions I have resorted to with success, as in both of them other methods had previously failed, or there had been a reproduction of a fresh crop of veins, which caused so much mental, if not physical distress, that I had no alternative but to comply with the patient's urgent request to attempt to relieve him. In planning this operation my object was to expose the veins of the cord by an incision, and then, as far as possible, to destroy them by means of the thermo-cautery. In this I was successful. Not only were the veins obliterated, but the operation was followed by the production of a

---

[*] "On the Radical Cure of Varicocele by the Galvanic Ecraseur."—*The Lancet*, July 17, 1880.

cicatrix so contractile as, so far, to prevent their re-appearance. I succeeded in obtaining, in the scar which resulted, the strong tendency to contraction which follows the healing of wounds inflicted by burning or scalding. In both instances the results I desired followed. In one, the inconveniences complained of entirely ceased after the operation, whilst in the other, though the local disease was cured, the mental apprehensions remained unrelieved, thus illustrating the observation of Paget to which I have already referred.

INDEX.

Abscess resulting from stricture, 145.
Aconite and quinine in urethral fever, 96.
Air, dilatation by means of, 86.
Albumen in the urine, 54.
Anæsthetics, 79, 303.
Aspiration, 103, 167.
Aspirator, use of the, 107, 355.
Atony of the bladder, 235.

Banks' cases, 40, 89, 172.
Belladonna, use of, 84.
Bell's view of syphilitic strictures, 136.
Bigelow's operation, 306.
Bladder, atony of the, 235.
    extra—peritoneal rupture of the, 321.
    female, foreign bodies in the, 188.
    inflammation of the, 225.
    injuries to the, 315.
    intra-peritoneal rupture of the, 320.
    irritable, 190, 233.
    male, foreign bodies in the, 179.
    puncture of the, 28, 323.
    rupture of the, 39, 315.
    sacculated, 200.
    sloughing of the, 371.
    stone in the, 270.
    tumours of the, 351.
    ulceration of the, 363.
    washing out the, 294.
Blood in the urine, 58.
Bougie, the filiform, 71.
Bougies, metallic, 76.
    oval, 119.
    whalebone, 16, 73.
Buckston Browne's tampon, 295.

Cadge's views on the formation of calculi, 244.

Calculi, causes of, 245.
    composition and shape, 251.
    early detection of, 255.
    formation of, 219, 239.
    impacted in the ureter, 265.
    impacted in the urethra, 283.
    physical constitution of, 239.
    size and shape of, 255.
    spontaneous fracture of, 248.
    table for the examination of, 253.
Calculous disorders, 262.
    treatment of, 285.
Calculus weighing 1,050 grains, 299.
Cantharides, tincture of, 236.
Catheter, passing the, 67.
Catheterism, 27, 66, 102, 317.
    forcible, 111, 217.
    prolonged, 66.
Catheters, 69.
Chancre of the meatus, 137.
Cheselden's operation, 288.
Chlorate of potash in cystitis, 235.
Chloroform, 79.
Circumcision, 373.
Cleanliness, importance of, 77.
Clover's aspirator, 230, 306.
Cock's operation, 110.
Colotomy, 368.
Contracture du col vésical, 43.
Copaiba and cubebs, 53.
Cystic oxide calculus, 299.
Cystitis, 225.
    treatment of, 227.
Cystotomy, 201.

Deposits, urinary, 55
Desormeaux's speculum, 13.
Dilatation, gradual, 70, 77, 80.
    continuous, 81.

Dittel's classification of stricture, 14.
Diverticula, 279.
Durham's plan of urethrotomy, 121.

Endoscope, the, 13.
Epispadias, 380.
Ergot and ergotine, 223, 236.
Extravasation of urine, 38, 147, 162.

Fæcal matter in the urine, 63.
False passages, 99.
Fasciæ, attachments of, 29.
Foreign bodies in the urethra and bladder, 179.
Frænum, deformities of the, 378.

Gauge, the, 78.
Gleet, 5.
　treatment of, 11.
Gonorrhœa, treatment of, 6.
Gunshot wounds of the bladder, 325.

Hæmaturia, 271, 335.
Hæmorrhage from the urethra, 98.
Hernia of the kidney, 328.
Holt's operation, 122.
Hypospadias, 380.

Incontinence of urine, 35, 204, 218.
Injections, use of, 5.
Instruments employed in the treatment of stricture, 69.
Inunction, 143.
Irrigation of the urethra, 11.

Jordan's (Furneaux) operation, 301.

Kidney, aspiration of the, 339.
　disorders of the, 95, 266.
　hernia of the, 336.
　injuries to the, 328.
　stone in the, 262.
　surgery of the, 337.

Litholapaxy, 306.
Litholysis, 309.
Litho-nephrotomy, 344.

Lithotomy, 286.
　supra-pubic, 313.
Lithotrite, use of the, as an extractor, 185.
Lithotrites, 304.
Lithotrity, 303.

Meatus, chancre of the, 137.
　deformities of the, 377.
　incision of the, 115.
Menzel's experiments, 318.
Mercury, use of, 141.
Microphone, the, 276.
Mucus and epithelium in the urine, 57.
Musculo-organic stricture, 118.

Napier's sound, 275.
Nephrectomy, 348.
Nephrotomy, 340.
Nitric acid as a solvent of débris, 309.
Nunn's view of syphilitic strictures, 136.

Otis's views, 24, 113.

Palpation of the rectum, 361.
Penis, amputation of the, 385.
Perinæal fistulæ, and their treatment, 127, 166.
　section, 125, 161.
Pessary catheters, 233.
Prepuce, contracted, 373.
Prostate, hypertrophy of the, 196, 209.
　treatment, 219.
　tumours of the, 351.
Pus in the urine, 59.
Puzey's case of nephrotomy, 340.

Quinine, injections of, 232.

Rectum, prolapse of the, 271.
Renal colic, 263.
Rest in the treatment of stricture, 65.
Retention of urine, 101, 209.
Rupture of the bladder, 315.
　of the ureter, 328.

Scar-tissue, 83.
Scirrhus and medullary cancer, 353.

Selection of cases, 112, 126.
Seminal ducts, opening of the, 31.
Septicæmia and pyæmia, 154.
Solvents of débris, 285, 309.
Sounding, 272.
  sources of error in, 279.
Sounds, varieties of, 275.
Spasm of the urethra, 21.
Spermatic fluid, 60.
Spermatorrhœa, 61.
Stammering with the urinary organs, 42.
Stimulants, the use of, 234.
Stone in the bladder, 270.
  in females, 312.
Stricture, causes of, 4.
  classification of, 14.
  consequences of, 35, 145.
  definition of, 4.
  impassable, 103.
  multiple, 15.
  nervous affections simulating, 41.
  position of, 9.
  symptoms of, 33.
  traumatic, 8.
  treatment of, 36, 65.
  varieties of, 17.
Subcutaneous surgery in the treatment of stricture, 133.
Suppositories, 233.
Suppression of urine, 96.
Supra-pubic incision into the bladder, 87, 218, 313.
Suspensory bandages, 390.
Syme's operation, 125.
Syphilitic strictures, 136.
  treatment of, 142.

Tapping the bladder, 108.

Thompson, Sir H., on the size of the urethra, 23.
Tumours, non-malignant, 356.

Umbilical urinary fistula, 385.
Urachus, patent, 384.
Ureter, injuries to the, 328.
Urethra, action of the, 180.
  calculi impacted in the, 283.
  curvature of arc, 25.
  dimensions of the, 23.
  female, dilatation of the, 312.
  hæmorrhage from the, 23, 98.
  injuries to the, 157.
  its relations to the rectum, 28.
  longitudinal wounds of the, 163.
  nervous and spasmodic affections of the, 41.
  rupture of the, 158.
  spasm of the, 21.
  surgical anatomy of the, 19.
Urethral abscess, 145.
  fever, 89.
Urethritis, granular, 35.
Urethrotome, the Author's, 117.
Urethrotomy, external, 125, 174.
  internal, 112.
  subcutaneous, 133.
Urine, deposits in the, 55.
  examination of the, 46.
  extravasation of, 38, 147, 162.
  retention of, 101, 209, 327.
  suppression of, 96.

Varicocele, treatment of, 388.
Villous growths, 357.

Watson's probe-pointed catheter, 103.
Wheelhouse's operation, 128.

www.ingramcontent.com/pod-product-compliance
Lightning Source LLC
Chambersburg PA
CBHW051736300426
44115CB00007B/588